Design for Profitability

Guidelines to Cost Effectively
Manage the Development Process
of Complex Products

Industrial Innovation Series

Series Editor

Adedeji B. Badiru

Air Force Institute of Technology (AFIT) – Dayton, Ohio

PUBLISHED TITLES

Project Management: Systems, Principles, and Applications, *Adedeji B. Badiru*

Project Management for the Oil and Gas Industry: A World System Approach, *Adedeji B. Badiru & Samuel O. Osisanya*

Project Management Simplified: A Step-by-Step Process, *Barbara Karten*

Quality Management in Construction Projects, *Abdul Razzak Rumane*

Quality Tools for Managing Construction Projects, *Abdul Razzak Rumane*

Social Responsibility: Failure Mode Effects and Analysis, *Holly Alison Duckworth & Rosemond Ann Moore*

Statistical Techniques for Project Control, *Adedeji B. Badiru & Tina Agustiady*

STEP Project Management: Guide for Science, Technology, and Engineering Projects, *Adedeji B. Badiru*

Sustainability: Utilizing Lean Six Sigma Techniques, *Tina Agustiady & Adedeji B. Badiru*

Systems Thinking: Coping with 21st Century Problems, *John Turner Boardman & Brian J. Sauser*

Techonomics: The Theory of Industrial Evolution, *H. Lee Martin*

Total Productive Maintenance: Strategies and Implementation Guide, *Tina Agustiady & Elizabeth A. Cudney*

Total Project Control: A Practitioner's Guide to Managing Projects as Investments, Second Edition, *Stephen A. Devaux*

Triple C Model of Project Management: Communication, Cooperation, Coordination, *Adedeji B. Badiru*

FORTHCOMING TITLES

3D Printing Handbook: Product Development for the Defense Industry, *Adedeji B. Badiru & Vhance V. Valencia*

Company Success in Manufacturing Organizations: A Holistic Systems Approach, *Ana M. Ferreras & Lesia L. Crumpton-Young*

Essentials of Engineering Leadership and Innovation, *Pamela McCauley-Bush & Lesia L. Crumpton-Young*

Handbook of Construction Management: Scope, Schedule, and Cost Control, *Abdul Razzak Rumane*

Handbook of Measurements: Benchmarks for Systems Accuracy and Precision, *Adedeji B. Badiru & LeeAnn Racz*

Introduction to Industrial Engineering, Second Edition, *Avraham Shtub & Yuval Cohen*

Manufacturing and Enterprise: An Integrated Systems Approach, *Adedeji B. Badiru, Oye Ibidapo-Obe & Babatunde J. Ayeni*

Project Management for Research: Tools and Techniques for Science and Technology, *Adedeji B. Badiru, Vhance V. Valencia & Christina Rusnock*

A Six Sigma Approach to Sustainability: Continual Improvement for Social Responsibility, *Holly Allison Duckworth & Andrea Hoffmeier Zimmerman*

Work Design: A Systematic Approach, *Adedeji B. Badiru*

Design for Profitability

Guidelines to Cost Effectively Manage the Development Process of Complex Products

Salah A.M. Elmoselhy

CRC Press
Taylor & Francis Group
Boca Raton London New York

CRC Press is an imprint of the
Taylor & Francis Group, an **informa** business

CRC Press
Taylor & Francis Group
6000 Broken Sound Parkway NW, Suite 300
Boca Raton, FL 33487-2742

First issued in paperback 2017

© 2016 by Taylor & Francis Group, LLC
CRC Press is an imprint of Taylor & Francis Group, an Informa business

No claim to original U.S. Government works

ISBN-13: 978-1-4987-2698-6 (hbk)
ISBN-13: 978-1-138-74871-2 (pbk)

Visit the Taylor & Francis Web site at
http://www.taylorandfrancis.com

and the CRC Press Web site at
http://www.crcpress.com

Contents

Foreword

by Jose L. Salmeron
Professor of Information Systems
University Pablo de Olavide, Seville (Spain, EU)

I was so glad when Salah Elmoselhy offered me the privilege of introducing the reader to this book. I really enjoyed reading it. The author has written a very interesting and comprehensive book about product design.

This book, written by an expert, contains everything that those involved in product design could possibly need to know, and it aims to be a repository for the current and cutting-edge applications of product design, an area with great demand in the market nowadays.

The product design field is growing and moving fast toward becoming a well-established discipline. This multidisciplinary field involves applying 2D/3D fabrication processes to produce commercial products and systems that entertain, enable, and inspire, and that change the way people all over the world live.

Currently, product designers are multi-faceted, with skills to build, integrate, and communicate ideas across different product areas, such as robotics or car design, fashion accessories, consumer products, and packaging. They combine business, entrepreneurship, technical, and design leadership to drive innovation and career success.

This book offers an empirical research–based design guideline that strikes a balance between complex product development time and performance, optimizing the added value in the product design process, especially in mobile robots.

Salah Elmoselhy presents a very efficient approach among the main proposals to understanding and managing the complex product design process. In addition, the book proposes a methodology to manage risks and hazards in the complex product design process in an efficient way.

As a result, this book can help novel designers to cost-effectively managing the complex product design process and enhance their competitiveness. Moreover, it can be a worthy tool for designers having a long career because of the innovative proposals included within.

Foreword

by Qingjin Peng, PhD, PEng
Professor
Department of Mechanical Engineering
University of Manitoba
Winnipeg, Canada

Profitability is essential for any product development. A product has to be developed meeting customer requirements on time and within budget. It is a challenge for any business to achieve a cost-effective solution in the process of product development. This book investigates variant factors that affect design for profitability to provide guidelines for readers to cost-effectively manage the product development process. The book provides research reviews in activities and strategies for success in the product development.

The book includes many relevant topics. Chapter 1 introduces product development modeling and process analysis. Methods in management and measurement of the product development process are described in Chapter 2. Chapter 3 discusses research methods to understand the product development process. Engineering design experiments are explained in Chapter 4. Chapter 5 discusses the design experiment description and outcomes. Statistical analysis results of the experiment outcomes are shown in Chapter 6. Chapter 7 discusses research findings' implications and conclusions. Remarks and future research directions are concluded in Chapter 8.

The author looks at mainly two research themes in the areas of process-oriented and product-oriented product development. The interaction of people, process, and facility is emphasized in the management of the product design and development process. Product attributes are discussed for customer requirements and product specifications. Different conflicting constraints are considered including time, cost, quality, and scope to determine the place where value is added in product development. Readers will find useful information in the book on internal aspects of product development to identify the most influential strategies and activities of product performance, and dimensions of the product development process.

This book also presents an example of a mobile robot design based on an empirical study of the robot design process to discuss key approaches and risks to manage the product development process. Readers will find approaches to achieve cost-effective solutions through improving the technical performance of their products to consequently improve their competitiveness in the market and their profitability.

At the end of the book, the author lists many directions for further research to introduce recent developments in the field including operational agility; design methods to embrace the uncertainty; design for sustainability; dynamic knowledge management and capabilities development; the integration of knowledge, technology road-mapping, and exchanging information; distributed development of products; the role of virtual networks and cloud computing strategy; integrating products and services; managing organizational culture change; and the user-centered approach. Big data, cloud manufacturing, and additive manufacturing have also been trends to push product development in a new direction. Readers can expect to see some details of these methods and developments in the field.

Foreword

by Huiyang Li, PhD
Assistant Professor
Systems Science and Industrial Engineering
Binghamton University (SUNY)
Binghamton, New York

Traditionally, the quality of product performance was considered a result of only the manufacturing process. This view, however, has changed. Based on recent research, the quality of product performance is fundamentally associated with product design. While the design process is the core of every new product development, it is not the whole story. Projects can also be manipulated by many factors that are not directly in the design process and thus are difficult to identify.

This research book proposes design guidelines to help product design managers and designers identify the parameters related to the design activities and strategies that significantly influence the product performance and consequently product success and profitability. Centered around two case studies in the structural design and construction engineering sector and robotics sector, the book first introduces readers to development process analysis, and approaches to manage and dimensions of the product development process. The book then gives an overview of the main research methodologies to understand the product development process, which engineers, designers, and managers may not be familiar with. In the following four chapters, details of two case studies of design management, from the design of questionnaire items to the final statistical analysis and discussion of the experimental results, are presented. Finally, future research directions in the field of product design are discussed.

This book addresses important issues in product design: factors influencing the design process, methods to identify them, and guidelines for future design. The field of product design is moving toward a well-established discipline. This growth relies upon the accumulation of research-based knowledge. This book contributes to this important knowledge base by presenting a comprehensive and in-depth literature review, and details of empirical studies. The focus on the empirical case studies, including the details, is very unique in published books in this field.

I am sure this book is going be helpful to researchers and students at various levels in the areas of product design, engineering design, engineering management, design management, and related fields. This book will also be very useful for engineers, designers, managers, and organizations who want to develop successful and profitable complex products through effective design management.

I wish all the best to Dr. Elmoselhy and the publisher. Above all, I wish all readers an insightful learning experience. Enjoy reading!

Preface

The goal of this book is to present a design guideline for complex products design and development companies to enable them to cost-effectively improve the technical performance of their products and consequently increase their competitiveness in the market and eventually improve their profitability.

This book explores engineering design management. As the name implies, this research area is interdisciplinary in nature, comprising engineering design, system design, project management, social sciences, business economics and strategic management perspectives. There are two research themes in this area: process-oriented and product-oriented. In process-oriented research the focus is on the product design and development process, particularly regarding the interaction among three aspects: people, process and facilities. This interaction is addressed from three perspectives. The first is that of managing knowledge and the socio-cognition throughout the product design and development process. The second is the process management perspective, which includes modelling and simulation of the product design and development process. Management of change and change propagation throughout the product design and development process is the third perspective. In product-oriented research, the focus is on the product itself, and computational design research represents a major perspective in this theme.

In designing complex products and integrated systems, industrial designers face the challenge of cost-effectively striking a balance between product development time and product performance attributes. To solve this dilemma, this book proposes a design guideline to effectively guide the product development process through identifying the root causes of product market success in product development activities and strategies. In addition, it has been reported that determining how and when value is added in product design and development is problematic. To solve this problem effectively, this book shows how and when value is added in product design and development through identifying statistically the design activities and strategies most and least correlated to product performance attributes. Hence, this book investigates statistically the correlation and causality between design process activities and strategies on the one hand and product performance attributes and consequently product market success on the other. It was found that an integrated design system has five subsystems: system design, project management, mechanical design, electronic design and software design. Across these five subsystems, the sets of design activities and strategies that were most positively and most negatively correlated with product success were empirically and statistically identified. The first set consists of (1) shifting the design complexity towards the software subsystem and (2) conducting the largest number of iterations in the software subsystem. It could significantly improve the product performance attributes by 20% and 25%, respectively (at the two-tailed 0.01 level); hence, it should receive the highest focus and the highest priority in assignment of resources. The second set consists of testing aggregately the software subsystem at the end of the project rather than adopting quick testing. It could have a significant negative influence on the product performance attributes by 26% (at the two-tailed 0.01 level); hence, it should receive the lowest priority in assignment of resources.

Moreover, this book proposes a hybrid approach to understanding and implementing the design process. By reviewing the relevant literature, it was found that there are six approaches to understanding and managing product design and development processes: (1) strategic and marketing-oriented perspective; (2) cost, accounting and financial-oriented perspective; (3) product development time-oriented perspective; (4) social sciences perspective; (5) technical-oriented perspective and (6) a hybrid approach of both strategic and technical approaches. It was also found that the hybrid approach is the most comprehensive and strikes a balance among the five pillars of product development: product attributes, product development time, product development cost, customer value and enterprise business goals.

This book thus presents a mobile robot design guideline based on an empirical study of the mobile robot design process. It also presents a hybrid lean–agile design paradigm for mobile robots. In addition, it points out key approaches and risks to consider in managing the product development process.

Acknowledgements

The author thanks the people of the Cambridge Engineering Design Centre, Cambridge University, for their help in accomplishing this work. He also thanks EPSRC and the Cambridge Overseas Trust for the support they provided for this work.

Author

 Salah A.M. Elmoselhy has been a PhD researcher in mechanical engineering working with the International Islamic University Malaysia (IIUM) and the Center for Sustainable Mobility at Virginia Polytechnic Institute and State University (Virginia Tech). He earned an MS degree in mechanical design and production engineering from Cairo University, and an MBA degree in international manufacturing systems from Maastricht School of Management (MSM). He has 10 years of industrial experience in CAD/CAM and robotized manufacturing systems. He recently became a researcher at the Engineering Department of the Fitzwilliam College of Cambridge University from which he earned a diploma of postgraduate studies in engineering design. He has authored/coauthored about 20 refereed publications, including ISI-indexed and SCOPUS-indexed journals and conference publications. His research appears in journals such as the *Journal of Manufacturing Systems* and *Journal of Mechanical Science and Technology*. He is an associate member of IMechE.

Introduction

Engineering design is a key phase in the product life cycle that can significantly influence the profitability of a product development company by influencing both the product performance and cost of the design process. Designs are integrated systems with several physical, functional and behavioural dimensions, and with several links between these dimensions (Pahl and Beitz, 1998). Because the commercial success of products significantly depends on the quality of product performance, inadequate management of the product design process leads to improper performance of products that can result in significant long-term business losses (Shenhar and Wideman, 1996). The potential influence of the product design and development activities and strategies on the market success of a product thus can be significant, particularly when developing complex products. Therefore, identifying and paying particular attention to the design activities and strategies that are the most influential in product performance can help to improve product performance cost-effectively; this consequently improves the value added to customers cost-effectively and hence improves the product design company's profitability in terms of improved product competitiveness and reduced design cost.

Business profitability is placed at the heart of industrial companies' strategic objectives. It could even be described as a 'common strategic objective' of all companies whatever their corporate strategic direction is, for example, growth, stability or retrenchment (Stevenson, 2002). Market pressure necessitates ever shorter development time, competitive quality of product performance and cost-effective product development and manufacturing. Because the revenue component of the industrial company's business profitability equation, profitability equals total revenue minus cost of goods sold, is determined entirely by market competitiveness and supply–demand equilibrium, the design and manufacturing operations can significantly influence and shape the cost component of this equation.

The major dimensions of this research area are process-oriented and product-oriented design management research. Process-oriented design management research investigates the management of the product design and development process from a process perspective. The process perspective investigates the steps of the process through which the product is designed and developed. People, process and facilities aspects are all employed to satisfy process-based performance measures. These measures include product development lead time and cost. They also include the study of robustness of the design process in an endeavor to make the design process less vulnerable to external effects (Chalupnik et al., 2007, 2008). The other way of investigating the product design and development process is through the product perspective, which emphasises the product performance attributes to make the designed and developed product more successful in the market. It measures the success of the design process solely on the basis of achieving the sought-after product performance attributes (Chalupnik et al., 2006, 2007). These two perspectives are reflected in the eight objectives that designers usually direct their attention to, either singly or in combination. The designs need to afford (1) manufacture, which includes design for manufacturability and design for assembly; (2) maintenance, which includes design for maintainability and design for serviceability; (3) retirement, which includes design for retirement; (4) sustainability, which includes environmentally conscious design and manufacture; (5) desired purposes, which includes systematic engineering design; (6) human use, which includes design for usability and ergonomics; (7) aesthetics, which includes industrial design and marketing and (8) improvement, which includes design for variety, design with modularity and product family design (Maier et al., 2007). Today, designers have started to shift from adopting only a product perspective or only a process perspective to synthesising both together in a self-designing product framework. That is, designers consider collectively technology that includes the scientific principles on which a design is based, geometry that represents the physical product and a process that includes the manufacture and use of a product. This integrated view is combined with the use of parametric design systems to allow parts and design components to morph into new forms as designs evolve.

The product design and development process consists of four phases: scoping, conceptual design, preliminary design and detailed design phase (Pahl and Beitz, 1998). The scoping phase is the stage in which the design plan is set and resource provisions and logistics are identified. The next phase is the conceptual design phase, in which the functional requirements of the product are identified and the design concepts of the required product are generated. A preliminary design of the required product is then generated, usually using computer-aided design (CAD) modelling. The final phase of the design process is generation of the detailed design of the product, which can include constructing and testing a to-scale prototype of the final product. The major dimensions of the product design and development process are the design people, process and facilities.

Each of these dimensions has characteristics that influence the product design and development process through either a design activity or strategy. In implementing projects, such as product design and development, there are four usually conflicting constraints: time, cost, quality and scope of the project. Usually, the project manager and team strive to satisfy these four constraints simultaneously. However, more often than not, they end up strictly satisfying three of these constraints at the expense of the fourth one.

In this book, *product attributes* refer to every feature of the product that adds value to the customer, for example, price, capacity, reliability and speed. These product attributes can be divided into two categories: (1) customer needs that are related to customer requirements and (2) product specifications that are related to engineering characteristics. The *design activity* represents every specific action that the design people take to achieve a specific minor goal in the design process. Making a three-dimensional CAD model of the product is an example of such a design activity. A plan of how to achieve specific objectives effectively in the design process is the *design strategy*. Adopting a modular design throughout the preliminary and detailed design phases is an example of the design strategy.

The principal product success criteria have five dimensions, each with measurable characteristics:

1. Internal project objectives
   - Meeting the schedule
   - Keeping within budget
   - Other resource constraints met
2. Technical performance objectives
   - Meeting functional performance
   - Meeting technical specifications and standards
3. Customer-oriented objectives
   - Favourable impact on customer's, that is, customers gain
   - Fulfilling a customer's needs
   - Solving a customer's problem
   - Improving the customer satisfaction index
4. Financial objectives
   - Revenue and profits enhanced
   - Larger market share generated
5. Strategic objectives
   - Will create new opportunities for the future
   - Will position customer competitively
   - Will create new market
   - Will assist in developing new technology
   - Has, or will add capabilities and competencies (Shenhar and Wideman, 1996)

The types of design projects vary and the relevance of success criteria varies accordingly. This book investigates two types of the four major types of design projects: established technology (classic-tech), mostly established (medium-tech), advanced (hi-tech), highly advanced or exploratory (super hi-tech). The two investigated types of design projects are type A, such as structural design and construction engineering projects, and type C, such as robotics and flexible manufacturing systems projects. These two types of projects are selected because the former represents a labour-intensive industry, whereas the latter represents a capital-intensive industry.

Product failure is the complementary part of product success. According to McMath and Forbes (1998), product failure is defined as (1) a product that was totally rejected by the market and ceased to exist or (2) a product that failed in market tests, resulting in a decision to abort its introduction. Therefore, only products that generated substantial positive financial results were defined as successes. Hence, critical success factors should be implemented to avoid product failure and meet the product success criteria.

According to Chan et al. (2004), there are five major groups of critical success factors in engineering projects: project-related factors, project procedures, project management actions, human-related factors and the external environment. More specifically, Baxter (1995) investigated critical success factors in new product development projects and found that the success factors in new product development that improve the likelihood of new product success are (1) strong market orientation in terms of significant benefits to users and value to customers (products that have this factor were five times as likely to succeed as those that had not); (2) early planning and specification in terms of sharp definition and precise specification of the product early in development (products that have this factor were three times as likely to succeed as those that had not); (3) company factors in terms of technical excellence, marketing excellence and technical and marketing synergy (products that have this factor were two and a half times as likely to succeed as those that had not). More recently, Belassi (1996) identified the critical success factors for design projects in the construction industry and in high technology and information technology systems industries in a ranked order as follows: (1) top management support, (2) project manager and team performance, (3) availability of resources, (4) preliminary estimates and (5) client consultation. In 2005, Clarkson and Eckert stated that success or failure of a product is significantly based on the interplay among abilities of the design team, their understanding of the customer's needs, the architecture of the product and the process by which it is realised.

Design processes are determined by multiple intertwined factors ranging from the characteristics of the product to the capabilities of the organisation. Thus, to have a better chance of competing successfully in world markets, products should be produced on time and within budget and meet customer and organisational requirements (The British Standards Institution, 2008). In particular, the major problem that leads to a decrease in sales despite implementation of statistical quality control techniques and quality improvements is failure to satisfy customers' requirements (Pyzdek, 2003). In addition, according to recent research, late introduction of new products has been shown to be the single largest contributor to the loss of companies' profit in the United Kingdom (Clarkson and Eckert, 2005). Accordingly, the design process in companies in the United Kingdom takes longer than it should and consequently is not managed cost-effectively.

Moreover, the statistics on business success worldwide show clear trends in product innovation. For every 10 ideas for new products, 3 will be developed, 1.3 will be launched, and only 1 will make any profit (Baxter, 1995). Therefore, managers in industry face the challenge of striking a balance between product development time and product performance attributes in the most cost-effective way. The available research literature identified this dilemma but has not as yet explored the root causes of product success in the product development activities and strategies. This research project is affiliated with the broad area of engineering design management science and it can help industrial product design and development managers in resolving this dilemma through investigating how product performance attributes can be influenced by design process activities and strategies. In particular, it aims at investigating the correlation and causality, if any, between design process activities and strategies on the one hand and the product performance attributes on the other. Therefore, the research pillars of this book are the five determinants shown in Figure I.1.

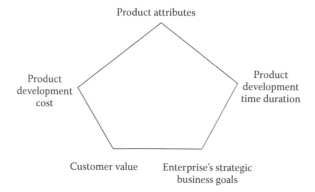

Figure I.1 Key determinants of complex products development process. (From Elmoselhy, S. A. M. (2015). "Empirical investigation of a hybrid lean-agile design paradigm for mobile robots." *Journal of Intelligent Systems*, 24(1).)

As Figure I.1 depicts, the key determinants of a complex product development process are the customer value and the enterprise's strategic business goals that collectively through the product development time and product development cost shape the product attributes. Therefore, to realise profitability in the product development process, these five pillars that are interconnected through the product development activities and strategies have to be harmoniously integrated. This book proposes a design guideline to effectively guide the product development process to achieve this harmonious integration.

In the product-oriented design research that is available to date, the emphasis is on minimising variations in performance caused by variations in uncontrollable noise parameters or in the design parameters (Chalupnik, 2006, 2007). In addition, Browning, Deyst, and Eppinger (2002) reported that in product design and development determining how and when value is added is problematic and thus is a research gap in the current literature. This book therefore aims to help to bridge this research gap. The proposed design guideline in this book helps product design managers and designers identify the design parameters themselves in terms of the design activities and strategies that significantly influence product performance attributes and consequently significantly influence the product's success in terms of the design company's profitability.

The book begins with product development process analysis and a modelling-based investigation. Chapter 2 presents dimensions of and approaches to managing the product development process. Research methodology to understand the product development process is presented in Chapter 3. Chapter 4 elucidates engineering design experiments. Description and outcome of design experiments follow in Chapter 5. Statistical analysis results on the outcome of the experiments are presented in Chapter 6. Chapter 7 demonstrates the implications and conclusions of the research findings. Finally, concluding remarks and future research directions are presented in Chapter 8.

The experimental results presented in this book are limited to complex products in the structural design and construction engineering and robotics sectors.

chapter one

Analysis and modelling of the product development process

For many years, product performance quality, an important determinant of the commercial success of a product, was considered the responsibility of the manufacturing team only. Recently, authors of some research papers have argued that product performance quality is fundamentally associated with product design. Accordingly they identified the following capabilities associated with robust product design: functional, aesthetic, technological and quality based (Clarkson et al., 2000; Swan et al., 2005). More recently, other authors have argued that product performance quality is fundamentally associated with the design process and thus have addressed the robustness of the design process (Browning et al., 2005; Chalupnik et al., 2006; Wynn et al., 2006).

1.1 Product versus process in developing complex products

Product-oriented research and process-oriented research are the two major dimensions in the product design management research area. Both of them attempt to maximise benefits and minimise risks to product design and development companies. What generates risks in developing products is the presence of uncertainties, which increase if the developed products are complex. Uncertainties are defined as 'things that are not known, or known only imprecisely' and comprise several categories, such as lack of knowledge, lack of definition, statistically characterised variables, known unknowns or unknown unknowns. They introduce various risks as well as opportunities (Hastings and McManus, 2004). In the context of engineering systems design, such risks may include failure in the execution of a task, longer than expected task duration, changes in product requirements or variations in available development resources (Hastings and McManus, 2004). Desired product attributes typically include reliability and robustness. Reliability is defined as 'probability that the system will work as specified under the specified operating conditions', whereas robustness is defined as 'ability of the system to work as specified in unexpectedly adverse environments' (Hastings and McManus, 2004), where the system can denote both process and product.

Both the process reliability and robustness are significantly influenced by the information on the behaviour of the process. For instance, timing of testing activities is based on proper understanding of the design process behaviour. On the one hand, if testing is carried out too late, discrepancies between desired and current states of the design may result in costly rework; on the other hand, testing too early may increase cost considerably without adding much valuable information (Chalupnik et al., 2006, 2007). Sections 1.1.1 and 1.1.2 discuss product- and process-oriented research in more detail.

1.1.1   Product-oriented research

The main goal in product-oriented design research is to design a product with inherent robustness through incorporation of robust design principles into the design process. The performance of the design process is therefore evaluated mainly on the basis of the robustness of the resulting product and from the wider perspective of its capability to reduce loss of quality. Product-oriented design research aims at improving the quality of products by means of statistical engineering as suggested by Genichi Taguchi (Taguchi and Clausing, 1990). He defines product or process optimisation as minimisation of loss in quality, which he perceives as loss to society (Goh, 1993). He defines robustness as a high signal-to-noise ratio, where the signal is what the system is trying to deliver, and the noise is the interference that degrades the signal (Taguchi and Clausing, 1990). Although in Taguchi's methods both process and product can be studied, traditionally the focus has been on the product (Chalupnik et al., 2006, 2007). Accordingly, design robustness of product or process can be one of two types. Minimising variation in performance caused by variations in uncontrollable noise parameters is called type I robust design. In contrast, type II robust design emphasises the uncontrollable variation in design parameters as a cause of variations in performance (Chen and Allen, 1996).

1.1.2   Process-oriented research

In process-oriented design research, the focus is on the performance of the design process itself, and the quality, reliability, cost, performance robustness and other attributes of the product are usually outside the scope of the analysis and are treated as exogenous variables. In measuring the performance of the process, several criteria, for example, based on the lead time or development cost, can be used. As the focus of process-oriented approaches to robustness significantly differs from that in product-oriented approaches, so does the choice of variables used. In design of a robust design process, the choice is determined in the first place by the modelling framework used, and, in the most general case, should take into account the use of the company's resources, constraints posed by the project's budget and time frame, as well as some organisational considerations. Because the engineering design process is a complex system, where complexity is not only technical, such as in product-oriented design research, but also, and more importantly, organisational in nature, its robustness can be analysed with respect to process, people and facilities. Improving performance of the design process can be achieved at the expense of either some deterioration in product performance or of extra outlay of resources to improve on these three aspects (Rosenhead, 1980). There are two ways of deploying extra resources to improve design process performance. The first is through an outlay of new development resources, such as human resource and facilities, as shown in Figure 1.1. The second is through spending resources on process improvement, which is a smarter approach; however, it takes more time and requires much more effort to be realised, as shown in Figure 1.2 (Repenning and Sterman, 2001).

Exclusion of interdependencies between the product-oriented and the process-oriented design research dimension in the analysis of product performance and process performance renders this analysis incomplete. An example is a robust design process in which mitigation of the influence of uncertainties is achieved by compromising product performance or quality. Despite significant short-term benefits that might thereby be accrued, such as delivering the project on time and within budget, a company's long-term reputation can be severely affected (Chalupnik et al., 2006, 2007).

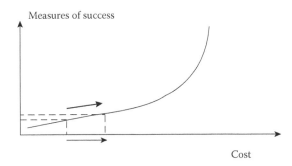

Figure 1.1 Process performance improvement obtained by means of outlay of new development resources. (From Repenning, N., Sterman, J. [2001]. "Nobody ever gets credit for fixing problems that never happened: Creating and sustaining process improvement." *California Management Review*, 43(4): 64–88.)

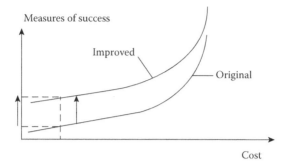

Figure 1.2 Process performance improvement obtained by means of spending resources on process improvement. (From Repenning, N., Sterman, J. [2001]. "Nobody ever gets credit for fixing problems that never happened: Creating and sustaining process improvement." *California Management Review*, 43(4): 64–88.)

1.2   Models and methods of conducting product design and development

Conducting product design and development involves (1) the establishment of appropriate structures for the design process; (2) the development and application of new design methods, techniques and procedures and (3) reflection on the nature and extent of design knowledge and its application to design problems (Cross, 1984). The methods in conducting product design and development range from the generation of better product design concepts (e.g. Pahl and Beitz, 1998) to the management of the design project (e.g. Baxter, 1995). This section presents the popular models and methods in product design and development research and implementation. They are design-focussed versus project-focussed methods, stage-based versus activity-based models of the design and development process, problem-oriented versus solution-oriented strategies for solving design problems and abstract versus procedural versus analytical models (Wynn, 2007).

1.2.1   Design-focussed versus project-focussed methods and models

A comprehensive classification of design management methods and models categorises them into design-focussed and project-focussed methods and models, which are discussed in Sections 1.2.1.1 and 1.2.1.2.

1.2.1.1   Design-focussed methods and models

Design-focussed methods and models support the generation of better products by application of perspective models and methods to the design process (Pahl and Beitz, 1998). Many relevant models and methods in the literature focussed on the design practice itself. French proposed a model that is based on design practice observed in industry and consists of the following four stages, as shown in Figure 1.3 (French, 1999):

- The process begins with the identification of a market need that is then analysed. This leads to an unambiguous problem statement and takes the form of a list of requirements that the product must fulfill.

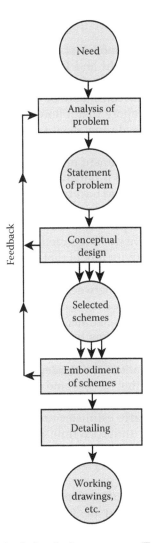

Figure 1.3 French's model of the design process. (From French, M. J. [1999]. *Conceptual Design for Engineers*. New York: Springer-Verlag.)

- During the conceptual design stage several concepts are generated, each representing a set of physical principles for solving the problem. These concepts are transformed into more concrete representations to allow assessment and comparison. The resulting concepts are evaluated and one or more are chosen to form the basis of the decisive solution.
- The chosen architecture is then solidified in the embodiment phase, where the abstract concept is transformed into a definitive layout.
- Finally, the remaining details are added to remove all ambiguity from the solution, leading to release of manufacturing instructions.

The most well-known model of the design-focussed phase-based models was proposed by Pahl and Beitz, as shown in Figure 1.4. Each of the four prescribed phases in that model consists of a list of working steps, which Pahl and Beitz consider to be the most useful strategic guidelines for design. They propose that following their prescribed steps will lead to more accurate scheduling, ensure that nothing essential is overlooked and result in design solutions that may be reused. In fact, Pahl and Beitz state that this document must be updated continuously as the design progresses, and that it is not always possible to draw clear borders between the phases or to avoid backtracking. Despite these apparent difficulties, they argue for the necessity of systematic methods to rationalise the design and production processes (Pahl and Beitz, 1998).

A different perspective was suggested by Evans (1959), who proposed a combined stage and activity model that explores the iterative nature of the design process. Noting that one of the most fundamental problems of design is to make trade-offs between many interdependent factors and variables, Evans argued that design cannot be achieved by following a sequential process. He demonstrated this using an example of bridge design, where the structure must be chosen to support the dead weight of the material, but the weight is not known until the structure has been defined. According to Evans such interdependencies are characteristic of design problems, a view that has become ubiquitous in modern thinking about design. Evans proposed that an iterative procedure be adopted to resolve such problems; early estimates are made and repeatedly refined as the design progresses, until such time as the mutually dependent variables are in accord. Based on this principle he proposed the prescriptive model for ship design shown in Figure 1.5.

1.2.1.2   *Project-focussed methods and models*

Project-focussed methods and models support and improve management of the design project, project portfolio or company (Hales, 2004). The project-focussed literature concentrates less on product design and more on product development, for example, cost-related activities such as

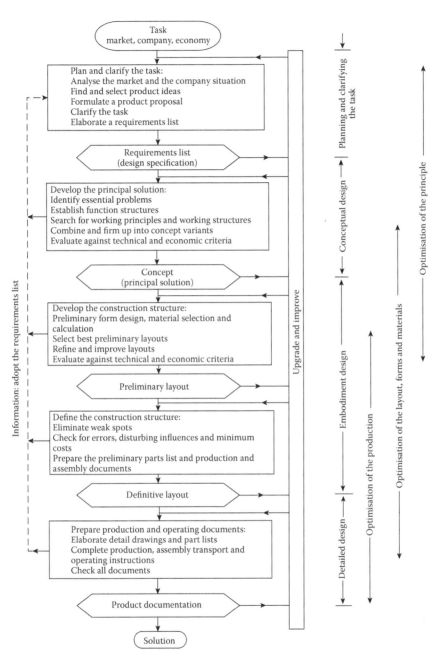

Figure 1.4 Steps of the planning and design process. (From Pahl, G., Beitz, W. [1998]. *Engineering Design: A Systematic Approach.* New York: Springer-Verlag.)

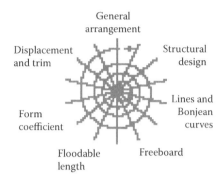

Figure 1.5 A summary of Evans' model of the ship design process. (From Evans, J. H. [1959]. "Basic Design Concepts." *Journal of the American Society of Naval Engineers*, November: 671–678.)

product planning and risk management (Baxter, 1995). Projects are influenced by a large number of factors that have little relation to the design process. Such influences vary from project to project. Management of these issues poses a challenge that Hales proposed is best resolved by promoting awareness of the influencing factors and their possible impacts on the project (Hales, 2004). He provided a comprehensive list of such influences at several different levels, including the macroeconomic, microeconomic and corporate scales, summarised in Figure 1.6. His method places the familiar stage-based view of the design process into its context within the project, company and market, and helps design managers to gain a broader perspective and to make informed decisions.

As another project-focussed method, Ulrich and Eppinger (2003) proposed meeting the following challenges in new product development by integrating personnel from a variety of backgrounds and perspectives:

- Recognising, understanding and managing product related trade-offs, such as weight versus manufacturing cost
- Working in an environment of constant change. As technologies and customer demands evolve and competitors introduce new products, there is a constant time pressure on all design and development activities
- Understanding the economics of product development from marketing through to manufacture and sales, so that a return can be made on initial investments

Pugh (1991a) argued that industry is concerned primarily with total design in terms of collaboration of personnel familiar with many different disciplines, both technical and nontechnical; for success in total design a company requires expertise in both technical engineering and

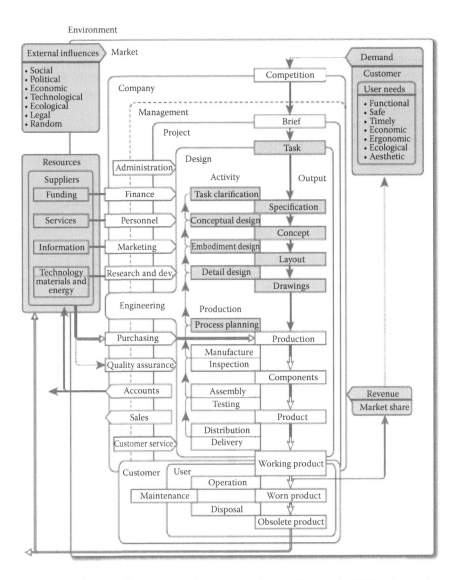

Figure 1.6 The design process in context. (From Hales, C. [2004]. *Managing Engineering Design.* New York: Springer-Verlag.)

engineering design, as well as many other nontechnical fields. This integration of disciplines requires that all participants have a common view of the total design activity with a minimum of misconceptions and can therefore contribute to a common goal effectively. Pugh proposed that visibility of operational structure is key to this common understanding, so that 'everyone can find out what people are doing and why'. He believed that a disciplined and structured approach is necessary to achieve this goal and proposed the model shown in overview in Figure 1.7. He presented the design process as a systematic and disciplined process that can be divided into six groups of interactive phases, generic to all kinds of design: market, specification, concept design, detail design, manufacture and sell.

Many project-focussed methods and models adopted in industry consider the design process as an ordered progression through a series of stages with gates between the stages. However, Cooper (1994) argued that

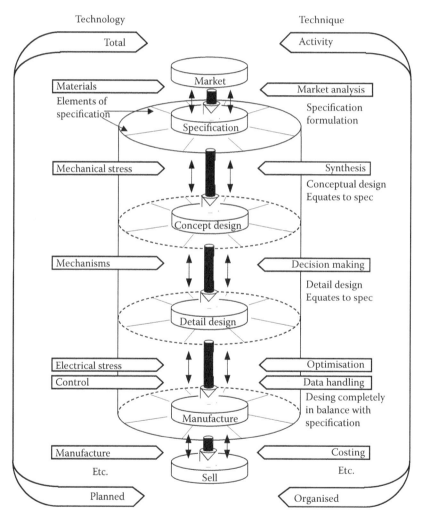

Figure 1.7 Total design activity model. (From Pugh, S. [1991a]. *Total Design: Integrated Methods for Successful Product Engineering.* Reading, MA: Addison-Wesley.)

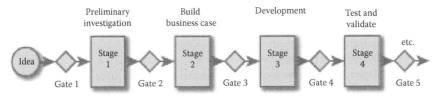

Today's stage-gate process.

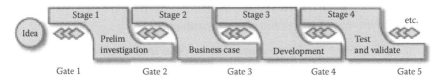

Tomorrow's 'third-generation process' with overlapping fluid stages and
'fuzzy' or conditional go decisions at gates.

Figure 1.8 Stage-gate structuring. (From Cooper, R. G. [1994]. "Third generation
new product processes." *Journal of Product Innovation Management*, 11: 3–14.)

there are many practical weaknesses to this form of gated process control. For instance, the system can be inefficient, as projects must wait at a
gate until all necessary activities have been completed. Therefore, Cooper
proposed that these systems should be made more fluid and adaptable.
In addition, he suggested that systems should incorporate 'fuzzy gates'
that should provide better management of the portfolio of products under
development, as shown in Figure 1.8.

1.2.2 Stage-based versus activity-based models of the design and development process

Blessing (1994) classified models of designing using the four categories
shown in Figure 1.9. This framework is based on the earlier theory of Hall

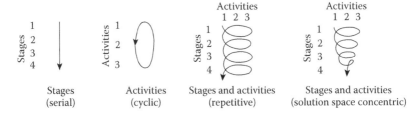

Figure 1.9 Stage- versus activity-based models. (From Blessing, L. T. M. [1994].
A Process-Based Approach to Computer-Supported Engineering Design. PhD thesis,
University of Twente, The Netherlands.)

(1962), who proposed a two-dimensional perspective of development projects in which the stage-based structure of the project life cycle lies orthogonal to an iterative problem-solving process that takes place within each stage in terms of activities.

Blessing refers to the essentially sequential and chronological structure of the project as a stage-based model, and the highly cyclical, rework-intensive, problem-solving activities as an activity-based model.

1.2.3   Problem-oriented versus solution-oriented strategies for solving design problems

Birmingham et al. (1997) used another scheme that categorises literature into either of the following two categories, according to the strategy the designer proposes be used to reach the design goal:

- Problem-oriented, in which emphasis is placed on abstraction and thorough analysis of the problem structure before generating a range of possible solutions
- Solution-oriented, in which an initial solution is proposed, analysed and repeatedly modified as the design space and requirements are explored together

1.2.4   Abstract versus procedural versus analytical models

Modelling the design process is employed as a method of mitigating risks inherent in engineering design. It facilitates gaining a deep insight into uncertainties, both those generic in design as well as those specific to the process. Once a better understanding of uncertainties is achieved, effective risk mitigation methods can be developed and deployed to improve process robustness. This scheme of categorisation classifies modelling of the design process into three categories: abstract, procedural and analytical models.

1.2.4.1   Abstract models

Abstract models are proposed to describe the design process at a high level of abstraction. Such literature is relevant to a broad range of situations but does not offer specific guidance useful for process improvement. A widely accepted abstract model is that proposed by Cross (2008), as shown in Figure 1.10.

Cross proposed a four-stage variant in which the designer first explores the ill-defined problem space before generating a concept solution. This is then evaluated against the goals, constraints and criteria of the design brief. The final step is to communicate the design specification either for manufacture or integration into a product or subsystem.

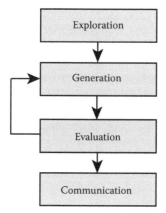

Figure 1.10 Cross' model of the design process. (From Cross, N. [2008]. *Engineering Design Methods: Strategies for Product Design.* Chichester: John Wiley & Sons.)

Because generation does not always result in a satisfactory solution, Cross includes a back track from the evaluation to the generation stage.

1.2.4.2   Procedural models

Procedural models are more concrete and less general in nature than abstract models because they incorporate a larger number of phases and focus on a specific aspect of the design process, industry sector and/or audience. Such literature is commonly categorised as follows (Finger and Dixon, 1989):

- Descriptive approaches result from investigation into actual design practice. Processes and procedures observed in industry form the basis of texts that are used primarily for teaching, training and research purposes. The abstract theories and models introduced in Section 1.2.4.1 are descriptive in nature.
- Prescriptive approaches are distillations of best practice intended to improve effectiveness or efficiency in some aspect of the design project. Such procedures are usually targeted towards a particular audience (e.g. student, design engineer or manager) and domain (e.g. industrial or mechanical design).

1.2.4.3   Analytical models

Analytical models are used to describe and evaluate specific situations and aim to provide insights and advice that may be operationalised in design processes. Such models consist of two parts: (1) a modelling framework used to describe aspects of a process and (2) techniques, procedures or computer tools that make use of the representation to support investigation or improvements to that process. Many modelling frameworks have been proposed as the basis of the analytical methods. Analytical models may be classified as follows (Browning and Ramasesh, 2007):

- Task network models represent processes as aggregations of tasks that may be tackled individually. These models assume that tasks are selected and attempted with the goal of driving the process towards completion. They often describe project stages together with the iterative, problem-solving steps within stages. With regards to the typology discussed previously, most task network models may thus be considered as combined stage- and activity-based in nature.
- System dynamics models view processes as work-processing systems whose behaviour is governed by the feedback and feed-forward of information about the current process state. In contrast to other approaches, system dynamics models of design processes are abstract and consist of few elements. The equations governing feedback are of critical importance in determining the dynamical behaviour of the model.

Each model type provides a different perspective of process structure and behaviour. Each is thus suited to modelling systems with certain characteristics. For example, task network models emphasise that system complexity arises from the structural connectivity of many simple elements, whereas system dynamics models assign an important role to the behaviour of individual components.

Task network models are more widely used in industry than system dynamics models. There are two major examples of task network models. The first are the design structure matrices (DSMs), one of the most important families of models in engineering design. In the original DSM method, the structure of design problems is displayed in terms of relationships among the technical parameters (Smith and Eppinger, 1997a). DSMs have two categories: static and time-based. Static DSMs have no time dimension and can be either people- or component-based DSM. Time-based DSMs can be either activity- or parameter-based DSM. Activity-based DSMs exemplify time-based DSMs, where the ordering of the rows and columns indicates a flow through time in the process (Browning, 2001), as shown in Figure 1.11.

The second example of task network models is signposting. Designing in the Signposting framework is viewed as identification and iterative refinement of parameters (Wynn et al., 2005). Parameters may denote any form of information that reflects changes in a process, and together with an associated confidence, which can represent the quality, accuracy or maturity of the parameter, are used to represent the state of the design

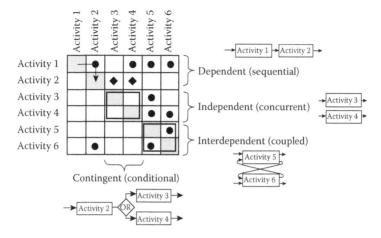

Figure 1.11 Example of DSM showing the four basic types of activities relationships. (From Browning, T. R. [2001]. "Applying the design structure matrix to system decomposition and integration problems: A review and new directions." *IEEE Transactions on Engineering Management*, 48(3): 292–306.)

process at any point (O'Donovan et al., 2002). Figure 1.12. shows the basic logic in the Signposting model is to signpost a designer to a next feasible (i.e. with all input parameters available) and useful (i.e. capable of increasing confidence) task. Additional criteria can be incorporated in the model to help a designer choose the best possible task.

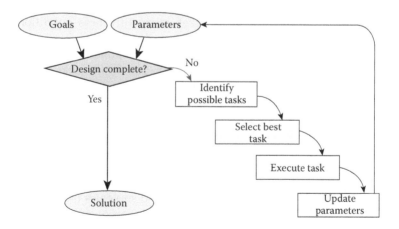

Figure 1.12 The Signposting model of the design process. (From Wynn, D. C., Clarkson, P. J., Eckert, C. M. [2005]. "A model-based approach to improve planning practice in collaborative aerospace design." In *Proceedings of ASME 2005, International Design Engineering Technical Conferences and Computers and Information in Engineering Conference*, Paper 2005-85297, pp. 1–12.)

1.3   Industry and profitability

The design process as presented in Section 1.2 is the core of every new product development project. However, it is not the whole story. Projects are always manipulated by a large number of factors that often have very little to do with the design process itself. Any design project is dependent on many of these external influences and thus should be considered in context; although certain types of project may have similar features, variations in context make each one unique. Hales (2004) highlighted a list of such influences at the macroeconomic, microeconomic and corporate scales, summarised in Figure 1.6. The figure places the design process into context within the project, company and market. Individual projects must often compete for limited resources within the company; from the market perspective they must integrate their processes with those of external contractors and suppliers. Individual companies may have responsibilities ranging from subsystem design to component manufacture; in either situation, successful integration of interorganisational processes is critical to prevent delays to the project. For example, the long lead time of an externally designed component means that errors in specification can be extremely costly. Other influencing factors are further removed from the project and cannot be managed or influenced directly. For example, changes in organisational structure, government legislation regarding the product or available manufacturing technology may cause a project to fail or be cancelled as a result of unforeseen economic factors. Because the revenue component of the industrial company's business profitability equation, profitability equals total revenue minus cost of goods sold, is determined entirely by the market competitiveness condition and supply–demand equilibrium, the design and manufacturing operations can significantly influence and shape the cost component of this equation.

In the product-oriented design research that is available to date, the emphasis is put on minimising variations in performance caused by variations in uncontrollable noise parameters or by variations in the design parameters (Chalupnik et al., 2006, 2007). In addition, Browning et al. (2002) reported that in product design and development determining how and when value is added is problematic and thus is a research gap in the current literature that this book aims to help bridge. The proposed design guideline in this book helps product design managers and designers identify the design parameters themselves in terms of the design activities and strategies that significantly influence product performance attributes and consequently the product's commercial success. Accordingly, the present work proposes that striking a balance among the five research pillars would improve a product design company's profitability; this is particularly evident when one considers that focussing on improving product performance would, beyond a certain level, lead to diminishing a company's profitability as indicated in Figure 1.13.

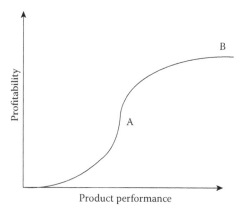

Figure 1.13 Product performance versus industrial company's profitability.

Such a specific level of product performance is the customer's acceptable level of quality, that is, the inflection point A on Figure 1.13, at which all customers' needs are fulfilled and beyond which the return on improving product performance will begin diminishing (Chalupnik et al., 2006). If the product performance is improved further than point A on Figure 1.13, by including the common desires among customers, the product performance should not exceed point B, which is the maximum profitable product performance at which the slope of the tangent to the curve becomes zero and no additional profitability can be attained by improving the product performance further (Rust et al., 1995). The issue tree that shows the overall rationale of this book in terms of how we can improve a company's profitability

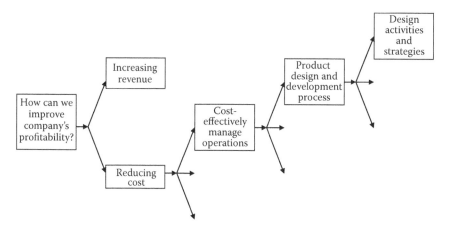

Figure 1.14 Issue tree of relevance of design activities and strategies to company's profitability.

and the relevance of the design activities and strategies is shown in Figure 1.14.

The issue tree in Figure 1.14 shows that a company's profitability can be improved by either increasing revenue or reducing cost. Cost reduction can be implemented through many branches including managing processes cost-effectively. These processes include product design and development, which can be managed cost-effectively by improving design activities and strategies. The design process can be understood better in the industrial context by reviewing the approaches used to understand and manage the product development process.

chapter two

Dimensions of and approaches to managing the product development process

According to authors of research papers that argue the design process significantly influences the quality of product performance, the design process consists of multiple intertwined factors: organisation-wide, product introduction process, design process, and individual-level (Browning, 1998). Accordingly, in the current customer-driven and demand-pull era, the first of these, organisation-wide factors, requires that the organisation needs to be managed strategically in terms of product portfolio management such that product development meets customer needs while at the same time coping with manufacturing constraints. Second, at the product introduction process level, all product development efforts should be integrated. Third, at the design process level, the designer should understand customer needs and market trends and translate them into a physical form that conforms to the company's perceived brand image (Cohen et al., 1996; Bayus, 1997; Erhun, 2007). Consequently, the product development activities and strategies should satisfy customer needs and expectations. Lastly, individuals should be supervised and inspired, effectively creating a paradigm shift in which managers become facilitators, and the mindsets of individuals are changed and integrated gradually. This view identified the importance of the design process to the quality of product performance but did not explore details of the relevant design activities and strategies.

Very recently, that view was supported by Paladino, who highlighted the importance of identifying the relationship between the success and the development process of new products as a research gap in the design management research area (Paladino, 2007). In addition, Slater stated that the alignment of the product development process and the strategic orientation of the firm along with addressing customer needs together can constitute the basis for successful technological innovation (Slater and Mohr, 2006).

Although the product development process is usually rational and based on gated stages – concept generation, product design, engineering analysis, process analysis and design and prototype production and testing, duration and progression of these stages usually cannot be pre-determined (Hastings and McManus, 2004). Such a probabilistic nature makes undertaking large-scale design projects difficult because of a combination of structural variations in terms of task connectivity patterns and uncertainties concerning task duration, rework likelihood, requirements of changes and resource availability (Hastings and McManus, 2004). Given these complexities and difficulties, if product development managers were to find a design guideline that advises on which design activities and strategies have a correlation and causal relationship with the success of the product, and accordingly assigned resources and priorities to these activities and strategies, they would rationally opt for this guideline. This chapter highlights the dimensions of this research gap.

2.1 Approaches to understanding and managing the product development process

The product design and development literature contains six approaches to understanding and managing the product design and development process: (1) strategic and marketing-oriented perspective; (2) cost, accounting and financial-oriented perspective; (3) product development time-oriented perspective; (4) social sciences perspective; (5) technical-oriented perspective and (6) hybrid approach of both strategic and technical approaches. Each of these approaches is discussed in this section.

2.1.1 Strategic, marketing-oriented and resources-based perspective

Improvement of product attributes can be a competitive advantage for any enterprise in terms of product portfolio management. In strategic management, the marketing decision consists of determining when to introduce the new product; whether the new product is a new version of an existing product, a sequence of products with overlapping life cycles or a new generation of a product and what the target performance level should be for the new product. Meanwhile, the strategic development decision consists of determining the allocation of the development time, resources and effort across development stages (Cohen et al., 1996; Bayus, 1997). These strategic decisions should help in realising the directional strategy of the firm, which could be growth directional strategy, in terms of increasing market share; stability directional strategy, in terms of maintaining the current level of profitability or retrenchment directional strategy, in terms of consolidation and striving to shift from losses to profitability by all viable means (Stevenson, 2002). Profit lies at the heart of all three of these directional strategies. Consequently, the relationship between the product development process, in terms of product development activities and strategies, and a company's strategic goals and profitability is significant. Several research papers and books explored the numerous aspects of the strategic and marketing-oriented perspective of product design and development.

In 1991, Clark and Fujimoto highlighted the impact of strategy, organisation and management on this critical component of business strategy. Using case studies and drawing on extensive field research the authors identified the strategies, practices and capabilities that improve lead time, productivity and product quality. They argued that the product development process is an important factor in determining the commercial success of the product and consequently that of the product development enterprise. Clark and Fujimoto (1992) explored product development and competitiveness of Japanese firms and found that what enabled them to develop and introduce more competitive products, faster and at lower development cost, are (1) capability in manufacturing, (2) rapid development and (3) efficient development.

To understand better how new product development projects can best be managed to respond to different conditions of uncertainty, Eisenhardt and Tabrizi (1995) made the point that extensive time spent planning is futile in highly turbulent and uncertain environments where speed is critical. Their study suggests that new product development strategy should be based on a thorough analysis of the characteristics of markets in which new products will be sold.

In 2001, Krishnan and Ulrich conducted a review of research in product development. They concluded that the critical success factors in engineering design are creative concept and configuration and performance optimisation. In addition, they found that there are eight product development decisions: (1) product strategy and planning, (2) product development organisation, (3) project management, (4) concept development, (5) supply chain design, (6) detailed design, (7) performance testing and validation and (8) production ramp-up and launch. In the same year, Tatikonda and Montoya-Weiss introduced an interdisciplinary view of innovation by integrating the operations and marketing perspectives of product development. Their findings showed that technological uncertainty moderates the relationship between organisational process factors and operational outcomes, and market and environmental uncertainty moderates the relationship between operational outcomes and market success.

In 2002, Debruyne et al. found that two thirds of new product launches meet reactions by competitors after their launch. Their study (Debruyne et al., 2002) found that competitors fail to respond to radical innovations and to new products that employ a niche strategy. They are more inclined to react to the introduction of new products that are supported by extensive communication and appear in high growth markets. In the same year,

Drucker argued that innovation success is more likely to result from the systematic pursuit of opportunities than from a flash of genius. He found that innovations based on new knowledge tend to have the greatest effect on the marketplace, but it often takes decades, and that ideas designed to revolutionise an industry rarely work.

In 2005, Chen et al. explored the relationship between speed-to-market and product success, asking whether faster is always better under different conditions of uncertainty. They found that speed-to-market is less important to new product success under conditions of low market uncertainty. Consequently, market, rather than technological uncertainty, moderates the speed–success relationship. To implement such a conclusion effectively, Erhun (2007) developed a process to facilitate decision making during new product introductions and transitions. The proposed process analysed the risks impacting a transition, identified a set of factors across departments tracking those risks, monitored the evolution of these factors over time and developed playbook mapping scenarios of risks and responses. Also in 2007, Paladino further facilitated such decision making by exploring the effect of both resource-based view strategy and market orientation strategy on firm innovation and new product success. The study found that design managers should focus on the provision of resources more than on the provision of customer value.

2.1.2   Cost, accounting and financial-oriented perspective

Cost-effective design is another approach in managing design and product development. Cost-oriented product development considers not only the development and application of product components but also the life cycle cost of the product, that is, cost of ownership, including design, production, operating, maintenance, reconfiguration and recycling. In this approach standard components as well as advanced technologies are candidates for cost reduction. Therefore, the focus that is adopted by numerous authors of research papers and books is not on cheap components but on the cost-effectiveness of the life cycle of the product. For instance, cost-oriented product development promotes cost savings in manufacturing processes and intelligent maintenance systems as well (Klaus et al., 2007). In 1992, Susman argued that companies that can most fully and rapidly integrate design for manufacturing concepts into their product development processes will generate quality, speed to market and product cost advantages that are critical to success in their business. In support of this approach, Hundal (1997) highlighted the importance of systematic designing and estimating costs during the design process – a time when it can be controlled most effectively.

Uncertainty in cost estimation has been a barrier to incorporating economics into the design process. Wilde (1992) demonstrated that, to achieve a higher quality product, an optimisation formulation is preferred to Taguchi's method. The study developed a methodology for formulation of multiattribute design optimisation problems by relating the decision variable, such as component geometry, which can effect improvement in manufacturing cost, to the performance attributes. To quantify beneficial trade-offs between cost and performance, in 1993 Thurston and Locascio evaluated the utility of the design process by constructing an objective function of subjective or 'unquantifiable' product attributes perceived by customer, and constrained such an objective function by functional relationships among design decisions, cost and performance. In 1994, Thurston and Locascio proposed a multiattribute utility decision-making method comprising two phases: preliminary design evaluation followed by fine-tuning for design optimisation.

Govindarajan (1993) married cost management methods to corporate strategy theories. He demonstrated how Activity-Based Costing provides useful insights into starting and measuring progress on a Kaizen exercise. Following this, Cooper and Slagmulder argued in 1997 that effective cost management must start at the design stage, as 90% of a product's costs are added in the design process. The primary cost management method to control cost during design is a combination of target costing, based on target selling price, and value engineering, based on achieving the required product functions at its target cost.

In 2000 Locascio developed an activity-based costing method to model manufacturing cycle time and cost. In that method, the factory operating costs are broken down by time, and cost is allocated to each product according to the processing it requires. That framework was implemented in a software program to aid designers in calculating manufacturing costs from limited design information. Several models attempt to quantify the 'manufacturability' of a design. The popular Boothroyd–Dewhurst index, for example, builds an estimate of design manufacturability relative to factors such as assembly complexity and number of parts (Boothroyd et al., 1991). Other models attempt to quantify design for X metrics to guide design decision making (Thurston and Locascio, 1994) or model trade-offs between design goals (Otto and Antonsson, 1991). These methods provide an assessment of the worth of the overall design in a way that remains generally applicable. In an endeavor to apply the principles of these methods, Wollherr and Buss proposed in 2001 the use of inexpensive standard hardware and software in place of high-cost commercial solutions to set up graphical virtual reality environments and simulations.

In 2002, Sandstr and Toivanen presented a management tool to connect design engineers to the economic objectives of a product development company. They suggested that the performance analysis of design engineers should be based on the balanced score card concept such that product designers do their work according to both target costs and the company's strategic goals. Yet, Andrew and Sirkin found in 2003 that most new products do not generate revenue because executives do not realise that the approach they take to commercialising a new product is as important as the innovation itself. They proposed three approaches to commercialising a new product and argued that selecting the most suitable one often yields two or three times the profit of the least optimal approach.

In 2007, Klaus et al. argued that cost accounting represents how cost originates and consequently described the organisation of cost management in view of its importance in product development. They found that cost unit accounting with differentiating overhead costing is central because it represents the common costing procedure in product-based industries. However, because of the limits of differentiating overhead costing for cost management, they presented more cause-based approaches such as process-costing and direct-costing. They also described the factors that influence manufacturing costs.

2.1.3   Time-oriented perspective

Product design and development time is the most apparent concern to design managers. Several research papers and books explored the numerous aspects of the time-oriented perspective of product design and development. Although high-tech design software can provide photo-realistic images ideal for the concept stages of product development, a physical three-dimensional prototype of a new product or component can speed up the approval process and can play a vital role in functionality testing, assembly trials and final design verification. Selective Laser Sintering (SLS) and Direct Metal Laser Sintering (DMLS) are two of the modern technologies that are recently used in this regard.

In 1987 Topping et al. presented a framework to integrate and manage the specifications with respect to both specification and execution time. They described how to generate software programming code automatically via a series of transformations of the specifications format that preserve correctness.

In 1996, Beesley found that time compression is a powerful source of competitive advantage that is relatively underutilised in UK businesses. The article explored the nature of time compression in relation to the fundamental principles of the supply chain and the concept of 'design for logistics' and proposed using time-based process mapping. Supporting this, Cohen et al. introduced in the same year a design process model that captures the trade-off between reduction of new product development cycle time and improvements in product performance. Their article determined the minimal speed of product improvement required for profitably undertaking new product development.

In 1997, Smith and Reinertsen presented numerous fast-to-market techniques including incremental innovation and emphasised that the fuzzy front end of the design process is the least expensive place to reduce cycle time. In line with this, Chen and Cheng proposed in 2000 that to provide a greater variety of complex products in shorter product development cycles while eliminating the hidden cost of part and process rework, product and process design activities should be integrated by using rapid prototyping and work-cell simulation. Supporting this, Jikar and Ragsdell proposed in 2005 a methodology for rapid product development by converting the functional requirements into product system, subsystem and piece part level and then selecting the best concept in terms of functionality and cost. Their article proposed using this in parallel with the Taguchi system of Quality Engineering as a strategy to predict warranty costs early in the design process. In support of that rapid product development view, Liou et al. in 2007 highlighted the importance of simulation and modelling and actual part building to time competence and recommended direct laser deposition to this end.

In 2001, Eppinger found that minimising design iterations is the key to equilibrate maximising a complex product's functionality with minimising waste of time and resources. He proposed the Design Structure Matrix (DSM) as a tool to represent design complexity in terms of design iterations needed to converge the design solution of the system parameters to conform to design constraints. In 2002, Browning et al. proposed that to realise an efficient and predictable design process, innovative design process architectures should be achieved and that DSM is helpful to this end. Very recently, Chalupnik et al. (2006, 2007) attempted to remedy drawbacks of the DSM by understanding design uncertainty and thereby design robustness through P3 Signposting modeller software. The P3 Signposting modeller software was introduced in 2006 by Wynn et al. and provides a parameter-driven task-based design process planner and more interactive modelling.

2.1.4   Social science perspective

Numerous research papers and books have explored the various aspects of the process of product design and development from a social science perspective. It has been adopted since 1995, when Bloch reported that the physical form or design of a product is an unquestioned determinant of its marketplace success. The article presented how the form of a product relates to consumers' psychological and behavioural responses and the strategic implications therefrom. In 1996, Nishiguchi's book set social bases for effective new product development and showed how to effectively manage the new product development process internally within the organisation and how to coordinate this with suppliers.

Veryzer (1998a) identified the key factors that affect customers' evaluations of a new product and found that they can improve the chances of making the right design decisions. These factors were familiar attributes of a product to customers and customers' uncertainty about a product's benefits and risks. Thus, Veryzer (1998b) found that a product's usability is important for customers' assessment of its quality and proposed mapped categories of product usability. Babbar et al. backed this in 2002 and found that alignment with customers is as important as a product's technical excellence for business success. Capitalising on those findings, Keates and Clarkson introduced in 2003 the concept of design exclusion and how addressing exclusion explicitly identifies and remedies many shortcomings in everyday products. Gemser et al. further backed this by reporting in 2006 that the three basic customer requirements are a product's usability, aesthetics, and functionality, and that sticking to these critical needs of the customer contributes significantly to product success.

Also in 2003, Yassine et al. found that information hiding leads to persistent recurrence of problems (termed the design churn effect) in the design process. Their article presented types of information flows, as well as a dynamic model of work transformation to derive conditions under which churn is observed, which consequently enables design managers to explore and compare the effects of improvement actions by managers. In addition, they proposed mitigation strategies to combat the design process churn. Joglekar and Ford (2005) agreed with this article that making myopic resource allocation decisions based on the observance of churn should be avoided. In an endeavour to help eliminate such information hiding in product development from a social science perspective, Lindemann in 2003 published a book that aimed to improve and evaluate tools and methods that support design by addressing individual thinking and acting and by addressing interaction between individuals. Very recently, Maier et al. (2008) explored, based on statistical analysis of empirical data, the correlations between the factors that influence communication in complex product development and identified the core factors among those.

2.1.5   *Technical-oriented perspective*

Several research papers and books adopted a technical-oriented perspective while looking through and analysing the product design and development process. The technical-oriented perspective indicates a specific engineering solution to a specific technical engineering problem. Therefore, the aim of this perspective is to find such specific engineering solutions. The research papers and books that address a specific technical engineering problem span virtually all types of technical engineering problems.

Smith and Eppinger (1997b) presented a model that enables design managers to determine the design subproblems that lead to design iteration and consequently dictate design process duration. They recommended mitigating strategies to accelerate the design process. To help solve the same problem, Ford and Sterman presented in 1998 a system dynamic model that enables managers to understand the dynamic concurrence relationships that constrain the sequencing of tasks as well as the effects of and interactions with resources (such as labour), project scope and project targets (such as delivery dates).

To effectively create design concepts to solve design problems, Terninko et al. presented in 1998 a method for structuring and focussing creativity and breakthrough thinking for problem solving. In 2004, Eris suggested a new design thinking model that illustrates the effective transformation of design requirements into design concepts and those concepts into design decisions and specifications as a question-driven process. Following that, axiomatic design was introduced, providing a helpful tool to identify the most suitable design process variables and parameters for the product functional requirements (Guenov and Barker, 2005). However, the axiomatic design approach assumes equal importance and correlation with product success for the full list of design activities from which the axiomatic design approach selects the most suitable ones for the product functional requirements.

In 2005, Esterman et al. found that effective management of reliability validation during product development and of system reliability is key to achieving superior time to market and life cycle quality, respectively. In support of that finding, Beiter et al. suggested in 2006 that to apply the Design for Manufacturing concept to complex software that controls highly technological products, the system should be defined and described.

Since 2000 numerous articles have emerged presenting robotics technical oriented issues. Hardt et al. suggested in 2000 design optimisation of a small, fast-walking, autonomous humanoid robot. They investigated nonlinear optimisation for a multilegged system. In 2005, Harkins et al. presented a research work to create an autonomous highly mobile water-resistant robot. Gonzalez De Santos, P. et al. found in 2007 that a better distribution of the legs around a robot's body can help decrease actuator size in the design procedure and reduce power consumption during walking as well, which is of vital importance in autonomous walking robots. In the same year, Castillo et al. described the use of a genetic algorithm (GA) for the problem of offline point-to-point autonomous mobile robot path planning. The problem consisted of generating 'valid' paths or trajectories, for a robot to use to move from a starting position to a destination across obstacles.

2.1.6 Hybrid perspective

The most comprehensive perspective in manipulating the various aspects of the product design and development process is the hybrid approach, which hybridises the other five perspectives to effectively manage the product design and development process in an integrated fashion. Several research articles and books that adopt and foster this hybrid perspective have been published since 1993.

Bakerjian et al. presented in 1993 a hierarchical model of cost variability throughout the product life cycle that replaces the traditional model of dichotomy of fixed and variable costs. They suggested adopting target costing at the front end of the design process. Following that, in 1994 Clausing introduced the term 'total quality development' and presented integration of concurrent engineering, Quality Function Deployment (QFD) and Taguchi's quality engineering into a total product development process. Capitalising on these advancements in this perspective, Baxter argued in 1995 that the secret to successful innovation is risk minimisation and identified the risk level at each product development phase. That book presented a structured framework for the management of innovation through systematic design and development methods that help to minimise the risk of nonconformance with functional requirements. Baxter (1995) found that successful application of systematic design methods requires an interdisciplinary and cross-disciplinary approach. A year later Huang presented the trade-off and integration between product development techniques, such as design for manufacture and assembly.

In 1999, Driva and Pawar showed that lack of available product design performance measures was identified as an industrial problem and highlighted five categories of performance measures: time, cost, quality, flexibility and unique parts and project organisation. Following those findings, Calantone and Di Benedetto (2000) found that firms respond to competition by either reducing new product development cycle time and/or improving product performance. They thus developed a model to equilibrate product performance and time to market by suggesting overlapping stages during which marketing, design and manufacturing engineering are jointly working on performance improvement. They found that if firms increase the length of time devoted to interfunctional integration, they can increase both the speed to market and product performance.

Also in 2000, Cristiano et al. investigated the use of quality function deployment industrially and reported improved cross-functional integration and better decision-making processes in the companies that implemented it. Following the same approach, Ulrich and Eppinger presented in 2003 a set of product development techniques that bring together the marketing, design and manufacturing functions of the enterprise. Availing of these advancements in this product development perspective,

Bayer presented a research paper that showed how quality can be built into product family architectures and how the quality of product family architectures can be assessed. That article introduced a common general process for design for quality accompanied by a general model. More recently, Cross (2008) published a book that aimed to help designers and design managers to develop a strategy that takes into account constraints including budget, time and material properties and addresses problem solving and optimisation.

Very recently, Browning and Ramasesh (2007) presented a survey of the product development process modelling literature that supports this product development perspective, focussing on activity network-based models. They categorised the activity network-based models into four major categories according to model purposes: visualisation, planning, execution and control and project development.

2.2   Dimensions of the interdisciplinary design process of complex products

The design and development of complex products are interdisciplinary tasks in nature. They necessitate a system design view to be in place to coordinate design activities effectively. In designing complex products, designers with combined knowledge in mechanical, electrical and software engineering will be able to devise more ideas for possible solutions and to evaluate better the feasibility of each idea. Recently the term 'Mechatronics' emerged, which is an interdisciplinary branch of mechanical, electronics and software engineering that is concerned with integrating electronics and mechanical engineering, usually in conjunction with software engineering, to create hybrid systems. Because designing complex products is usually a specific mission with specific objectives to be achieved within a specific time frame, the design and development of complex products can be considered a project. Therefore, considering and implementing the project management philosophy and techniques are key to successful design and development of complex products. In addition, complex products usually combine various subsystems such as mechanical, electronics and software. Therefore, coordinating design activities and strategies and managing interfaces between these subsystems are decisive factors behind success or failure of design and development of complex products (Tomatis et al., 2002).

2.2.1   System design dimension

The system design addresses the interrelationship between the various parts of the process and/or of the product. Each system and subsystem has a boundary, limits and information of the whole, as each subsystem is related to the whole. The whole has no effective meaning without its inter-related parts and the parts have no effective meaning without the whole to which they are related. Hence, system design aims to make the whole greater than the sum of its parts. To realise a successful system design, design strategies and activities can be utilised.

Modular design recently received the attention of researchers as an effective tool to achieve successful system design. Modular design implies splitting the overall system into modules that interface with each other. A few advantages of modular design include speeding up the design development process, facilitating error-tracing and maintainability and minimising impact of design iterations (Smith and Reinertsen, 1997; Clark and Baldwin, 2000). Modularity can range from product design, to manu-facturing processes, to end uses of a product (Gershenson and Prasad, 1997). The systematic design approach is another helpful design strategy that suggests following specific steps to achieve an effective design, such as Design Function Deployment (Sivaloganathan et al., 2000). The Design Function Deployment approach consists of seven steps: (1) establish-ing requirements and specifications based on customers' demands and wishes, (2) proposing conceptual solutions, (3) developing embodiment design, (4) developing detail design, (5) selecting material and manufac-turing processes, (6) generating production plans and (7) selecting the optimal design solution (Sivaloganathan et al., 2000). Another example of this systematic design approach is Pugh's concept selection process to evaluate design concepts that suggests the following steps: (1) clarifying the design problem; (2) identifying requirements specification and design constraints; (3) sorting the requirements specification in a ranked order; (4) identifying functional requirements, for example, using an axiomatic design methodology, and identifying core design values; (5) constructing a table of design options based on brainstorming; (6) making a concept evaluation, for example, based on a weighted criteria evaluation matrix and (7) making an overall strategic decision of the conceptual design (Pugh, 1991a,b; Pahl and Beitz, 1998). In the systematic design approach it is recommended that the number of investigated design concepts should not exceed five and should not be fewer than three (Wallace and Clarkson, 1999). Because some steps of the systematic design approach, such as the sixth step in the design options evaluation matrix, that is, making con-cept evaluation, are subjective in nature, the quality of performing the systematic design steps, such as getting clear customer requirements, is important to reducing the possibilities of making wrong design decisions (Sivaloganathan et al., 2000).

Error-proof is another design and development strategy that can help to realise a successful system design (Chao et al., 2005). In a mechanical subsystem, constructing a three-dimensional (3D) computer-aided design

(CAD) model of the system assembly drawing is a real-world example of the implementation of the error-proof strategy to minimise waste of time and resources due to avoidable mistakes. In addition, a 3D model of mechanical subsystem components can be used to check fittings of each component with each other, that is, concerning eccentricity and fits and tolerance, and to check conflict between motions of movable parts. Life-size 3D representation of a product using a 3D CAD model is another alternative that can be used in industry to speed up the product development cycle by reducing the need for physical prototypes, thus preventing error and saving time and money. In an electronics subsystem, error-proof strategies include mounting components on the printed circuit board in the following order: first mounting smallest components, second mounting largest components such as large capacitors and finally mounting the most heat-sensitive components such as transistors. In software subsystem, error-proof strategies include adopting parameterisation in software code implementation, that is, avoiding inconsistent change of value of a system parameter that is used in various places within the code.

Maximising functionality of the final concept of the product while minimising waste of time and resources is another concern that puzzled researchers in this area for years. Recently, Eppinger (2001) suggested that this balance can be achieved by minimising the design iterations. Adopting top-down system structure decomposition can also contribute to the success of the system design and can be realised by identifying the product functional requirements through analysing the functional structure of the product and consequently mapping it to the requirements specification (Bernard, 1999).

In the robotics and flexible manufacturing systems industrial sector, striking a balance between fast response on the one hand and stability, accuracy and payload fulfilment of the robot on the other is key to successful system design and eventually to satisfactory robot attributes and performance (Yavuz, 2007; Fanuc Robotics, 2008). However, such a balance should not be realised at the expense of the product reliability, maintainability and safety. The major characteristics that are usually ignored during the search for overall performance improvement are reliability, maintainability and safety (Pyzdek, 2003). Product reliability, which means the ability of the product to perform its required functions under stated conditions for a specified product service time, is significant to the success of the design of the product as a system (Coulibaly et al., 2008; Prabhakar Murthy et al., 2008). Also, maintainability of the final product in the design process, in terms of the ease with which maintenance of the product functional units can be performed, is key to a successful system design (Coulibaly et al., 2008). Safety of use of the final product in the design process, in terms of accident prevention, risk identification and risk control and/or minimisation, is important as well in this context

(Coulibaly et al., 2008). Finally, prediction of risk scenarios, for example inappropriate lighting conditions or missing a junction on the route, and preparation of action plans accordingly, for example implementing an error-recovery tactic, improves performance robustness of the product as a system (Rostami et al., 2005).

2.2.2   Project management dimension

Project management is a cornerstone of successful product design and development. The dimension of project management should be considered carefully in the product design and development process activities and strategies. It reflects the interaction of human resources, facilities, budget and strategic goals. As a helpful design activity that the design and development project manager should consider, visiting, that is, looking at and reviewing similar designs that solved similar design problems, can help in realising a successful product design (Dahl et al., 2001). In addition, conducting a design research activity, such as research and/or Internet literature review search to become acquainted with the state of the art of the design of the required product, can help in realising a successful product design (Dahl et al., 2001).

Adopting cross fertilisation of design concepts among the design team, such as sharing ideas and/or modifying each other's ideas, is a design and development strategy that should be adopted while managing the design process because it increases the likelihood of achieving an effective design concept (Amon et al., 1995). This strategy can be more effective if the design team is multidisciplinary such that each of the design team members is aware of more than one relevant discipline such as material science, design development approaches, leadership skills, and so forth and is more aware of the overlap and intersections between these disciplines and approaches to minimise the risk of errors (Amon et al., 1995).

The product design and development team have further aspects that can influence the product design and development process. Adopting live meetings of the design team in the conceptual and preliminary design phases and e-mails in the detailed design phase as the principal communication method, to strike a balance between the synergy of face-to-face meetings on one hand and the time effectiveness and message delivery effectiveness of the electronic means of communication on the other hand, can help in achieving successful design project (Court, 1998). In addition, the more the design team meets up collectively in the conceptual design phase than in detailed design phase, the greater the likelihood of achieving a more effective design concept (Court, 1998). The outcome of the design discussions among the design team is helpful in this context and should be documented because it is a valuable means of design concept improvement (Pahl and Beitz, 1998). Moreover, having experienced designers among the design team can help the team to better conceive the assumptions and constraints accompanied by similar designs and to make the design process more foolproof (Court, 1998). This can particularly help the design manager to redeploy the human resources of the design team to cope with fluctuations in the workload of project activities, which is a helpful strategy in managing a design project (Browning and Eppinger, 2002; Mahmoud-Jouini et al., 2004). Moreover, empowering and authorising design subteams to make tactical decisions without need to refer them to the team leader can help in achieving a successful design project (Krishnan, 1998; Pearce, 1999).

Resources are an ongoing concern to the project manager, and the design and development project manager is no exception. Having resource provisions for unforeseen problems can help the design and development project manager to minimise vulnerability of the design development process to the influence of external factors (Browning, et al., 2005). The internal factors in the design process, such as design mistakes, also influence the success of the design and development project. A modular testing strategy, that is, testing the deliverables between the system submodules as a way of verifying conformance of these submodules to the conceptual functional requirements, can help to detect mistakes as early as possible and consequently to minimise their impact (Lévárdy and Browning, 2005). Testable design interdeliverables within and among system modules based on project milestones can help as well in detecting mistakes as early as possible and consequently in minimising their impact (Huang, 2000; Lévárdy and Browning, 2005). In addition, modular deliverables and testing, that is, testing deliverables of each module, rather than subsystem deliverables and testing, can further help in minimising the impact of mistakes (Clark and Baldwin, 2000).

Managing the design and development project to meet the project delivery deadline, given the limited amount of available resources, is another ongoing concern to the design and development project manager. Usually, the design project manager is more interested in meeting project delivery deadlines than in adhering to specified costs to achieve the project objectives (Mahmoud-Jouini et al., 2004; Clarkson and Eckert, 2005). Therefore, the design project manager would be more interested in the critical path, the longest sequence of time-constrained design tasks in the project plan, than in the critical chain, the longest sequence of resource-constrained design tasks in the project plan (Leach, 2000). In addition, the design project manager would usually make every effort to help meet the project delivery deadline including adopting both sequential and concurrent design activities according to the nature of each task and the preceding relationships between them, to maximise utilisation of resources and shorten the design development time (Browning and Eppinger, 2002).

2.2.3   *Mechanical subsystem design dimension*

Mechanical engineering is an engineering discipline that applies principles of physics to analysis, design, manufacturing and maintenance of mechanical systems. As the most apparently tangible subsystem of any integrated design system, the mechanical subsystem acts as the skeleton of the product that deals with physical external loads. A major aspect in designing such a skeleton is to consider its manufacturability. Manufacturability includes order of manufacturing and assembly of system components; for example, by having one rivet to connect two members of the structure, these two members can still be rotated. Considering manufacturability starting from the conceptual design phase through the detail design phase in the design process can minimise waste of time and resources and can eliminate unnecessary iterations. In addition, checking the accuracy of manufacturing and assembly of the final design prototype can help avoid unexpected failure due to manufacturing defects and/or assembly mistakes (Pahl and Beitz, 1998).

2.2.4   *Electronics subsystem design dimension*

Electronics engineering design deals with the behaviour and effects of electrons, such as in transistors, and with electronic devices and systems. The objective of the electronics design subteam is to design, build and test the interface electronic circuits that will be the interface between the interdisciplinary system's transducers, which provide information on the 'real world', and the interdisciplinary system's microprocessor. The interface electronics circuit design process includes the following stages:

- Definition of circuit functions
- Circuit design and calculation of electronic components' values
- Circuit documentation
- Circuit construction
- Circuit test (O'Dell, 1988; Cambridge University, Engineering Department [CUED], 2007a)

   The electronics subsystem acts as the link between the mechanical and the software subsystem in any interdisciplinary design system. In addition, it provides the physical control system for the mechanical subsystem. Moreover, its reliability, which is generally greater than that of its mechanical subsystem counterparts, can be a prominent advantage of the electronics subsystem (Ascher, 2007).

2.2.5   Software subsystem design dimension

The essence of flexibility and major source of robustness in the interdisciplinary design system is the software subsystem. Software engineering is the application of systematic techniques and procedures to the development, operation and maintenance of software. Software design encompasses three steps: (1) software design, that is, data, architectural and procedural design; (2) code design, that is, program modules and (3) software code testing, to integrate and validate the software (CUED, 2007a; Sommerville, 2007). Recent advances in software code implementation techniques in the form of object-oriented programming further support the prominent advantages of software engineering. Many software designers prefer object-oriented to structured programming code implementation because of the effectiveness of the former to achieve a better design because functions and subroutines are less effective in terms of reusability, scalability and manageability. In addition, object-oriented programming code implementation offers the advantage of modularisation of classes (Rob, 2004). In addition, many software designers aim at implementing explanatory annotation of the software code for purposes of software code debugging and maintenance (Wasserman et al., 1990). To facilitate software code debugging, maintainability and/or scalability of the final product concept, a parameterisation design strategy, that is, changing the value of one parameter to change the values of many related parameters accordingly, is recommended (Loughran and Rashid, 2004). To strike a balance between minimising the cost of testing and detecting mistakes as early as possible in software code implementation, quick testing of interdeliverables between the modules of the software draft code, and extensive testing of the overall software draft code on a prototype printed circuit board or on an equivalent facility are recommended (Lévárdy and Browning, 2005).

2.3   Recommended approach to understanding and managing the product development process

This book adopts hybridisation of both product-oriented and process-oriented design research perspectives to cost-effectively improve product performance attributes and consequently to make the product more successful in the market, and eventually to improve design companies' profitability. In addition, the adopted approach to understanding and managing the product development process is the hybrid approach, as it represents the most comprehensive approach in this context.

Based on the results presented, adopting both lean robot design activities and strategies and agile robot design activities and strategies together in the mobile robot design process is proved in this book to be practically valid. In addition, the design experiment has proved that both lean and agile mobile robot design activities and strategies are correlated with and have significant influence on improving mobile robot performance. It has also been found that there are mobile robot design activities that have attributes of both the lean and agile design paradigms. For instance, evaluation of design concepts exhibits attributes of both lean and agile design paradigms. This further supports the practical validity of adopting both lean and agile mobile robot design activities and strategies together in the mobile robot design process. Therefore, this book proposes a hybrid lean–agile mobile robot design paradigm in which both lean and agile mobile robot design activities and strategies are adopted in the mobile robot design process, which thereby benefits from the attributes of both paradigms.

The proposed mobile robot hybrid lean–agile design pillars comprise (1) adopting the most effective lean design strategies, such as considering reliability of the mobile robot in the design process in terms of its ability to perform its required functions under stated conditions for a specified service time; (2) adopting the most effective agile design strategies, such as having designs that are less vulnerable to failure modes and less exposed and less sensitive to uncontrollable external factors by shifting complexity to the software rather than to the mechanical subsystem; (3) adopting the most effective lean design activities, such as testable design interdeliverables within and among system modules based on project milestones to detect mistakes as early as possible and to minimise their impact on the successful completion of the design project; (4) adopting the most effective agile design activities, such as having iterations in the software rather than in the mechanical subsystem to achieve a shorter development time; (5) adopting a three-phase hybrid lean–agile risk management action plan that helps in integrating mobile robot design activities and strategies to minimise risk in the mobile robot design process and (6) adopting mobile robot design functional strategy in terms of standard components, modular design, unified architecture of mobile robot chassis and frame parts and concurrent engineering in the design process.

The time frame of the proposed three-phase hybrid lean–agile risk management action plan is as follows: (1) before the beginning of the mobile robot design process in which strengths, weaknesses, opportunities and threats (SWOT) analysis is conducted; (2) during the mobile robot design process in which the design team proves the value of the design concept to stakeholders at the end of each design phase, ensuring that the mobile robot satisfies stakeholders, fits its intended purpose, is of a quality to last its design lifetime and can be made at an acceptable cost; (3) after the end of the mobile robot design process in which failure modes and effects analysis (FMEA) is conducted and ultimately the models of mobile robots that fall short of the set target are destroyed as soon as this appears. This approach to managing risk in product design process is expected to help significantly in realising the sought harmonious integration between the product development activities and strategies.

Key challenges for the hybrid lean–agile mobile robot design paradigm comprise cost-effectively striking a balance between robot quality and short duration of the design process. In addition, misreporting cost savings can hurt the credibility of the hybrid lean–agile mobile robot design paradigm practitioners. Moreover, lean savings are indeed a long-term proposition. As expected, much pressure is imposed on the implementers of the hybrid lean–agile mobile robot design paradigm to show immediate savings in terms of financial indicators such as the Return on Investment (ROI) to top management. Yet, reduction in defects and reduced cycle times are all areas that will continue to produce savings long after the term of ROI has run out. The implementation of the hybrid lean–agile mobile robot design paradigm thus needs a champion to lead the change cost-effectively and leverage it.

chapter three

Research methodologies to understand the product development process

Research methodology is central to the rigour and quality of research. This chapter presents the research methodologies that are used to understand the product design and development process and accordingly the research methodology adopted in this book. After a description of the three major relevant research methodologies used to understand the product development process, the particular research methodologies are presented. Next, other classifications of research methodologies to understand the product development process are described, followed by collected data types. Next, sampling and statistical errors are considered. Data collection methods, research validity methods and research verification methods are then considered. Finally, the chapter presents the adopted research methodology.

3.1 Research methodologies to understand the product development process

Three major research methodologies are generally used to understand the product design and development process: exploratory, descriptive study, and correlational study (Sekaran, 2003). Sections 3.1.1 to 3.1.3 elaborate on each of these three methodologies.

3.1.1 Exploratory study/causal study/experimental research

In an exploratory study, also called a causal study or experimental research, researchers attempt to establish cause-and-effect relationships among the controllable independent variables of the study – the cause and the measurable dependent variables – that is, measuring the effect on the dependent variables. This is usually done while the cause, that is, the independent variable, is under the control of the experimenter. The independent variables can be under full or quasi-control of the experimenter in the experimental settings (Cook and Campbell, 1979a). Sections 3.1.1.1 to 3.1.1.4

present the four types of exploratory study: true experimental study, quasi-experimental study, case study and action research.

3.1.1.1   True experimental study

A true experiment, mostly thought of as a laboratory study, is defined as one in which an effort is made to impose control over all other variables except the one under study. It is often easier to impose this sort of control in a laboratory setting. In order to better understand the nature of the experiment, the following terms are defined:

1. The experimental or treatment group is largely the group that receives the experimental treatment and manipulation.
2. The control group is used to produce comparisons based on the variable under study. It is a subset of the treatment group. It receives the treatment of interest that is deliberately withheld or manipulated to the rest of the treatment group to provide a baseline performance with which to compare the treatment group's performance.
3. The dependent variable is the variable that is measured in a study. The experimenter does not control this variable.
4. The independent variable is the variable that the experimenter manipulates in a study to examine its influence on the dependent variable.

In a true experiment, every experiment must have at least two groups: a treatment and a control group. Each group will receive a level of the independent variable. The dependent variable will be measured to determine if the independent variable has an effect. As stated previously, the control group will provide the experimenter with a baseline for comparison.

3.1.1.2   Quasi-(natural) experimental study

Quasi-experiments are very similar to true experiments but use naturally formed or preexisting groups. In this type of study, there can be many variables that are outside of the experimenter's control and could account for differences in the dependent measures. Thus, with quasi-experimental designs, the experimenter must be careful in making statements of causality.

Laboratory experimental settings increase the ability of the researcher to control the design environment and to gather many data rapidly by excluding some variables. Laboratory experiments have a great deal of internal validity in comparison with natural experiments when the measurement of some variables is affected by other variables. However, laboratory experiments lack external validity when the results of the experiment are applied to a real-world situation. Therefore, natural experiments that intrinsically have external validity can be as accurate as the laboratory experiments if a researcher could measure variables as independently as possible (Robson, 2002; Sekaran, 2003).

3.1.1.3   Case study

Rather than using large samples and following a rigid protocol to examine a limited number of variables, case study methods involve an in-depth, longitudinal examination of a single instance or event: a case. They provide a systematic way of looking at events, collecting data, analysing information and reporting the results. As a result, a researcher may gain a sharpened understanding of why the instance happened as it did. Case studies lend themselves to both generating and testing hypotheses.

A case study can also be defined as an empirical inquiry that investigates a phenomenon within its real-life context. Case study research can contain single or multiple case studies and can include quantitative evidence. It is situated somewhere between data collection techniques and theoretical propositions. Case studies should not be confused with qualitative research and they can be based on any mix of quantitative and qualitative evidence. Quantitative case-study data can provide the statistical framework for making inferences (Yin, 2002).

The usual basis for case study methods is deductive research and includes the following steps: (1) determining the research questions, (2) selecting the cases and determining the data-gathering and analysis techniques, (3) collecting data in the field, (4) evaluating and analysing the data and (5) preparing a report (Soy, 1997).

3.1.1.4   Action research

Action research is a reflective process of progressive problem solving, that is, a spiral of steps each of which is based on fact finding about the result of its preceding step. This process is led by individuals working with others in teams to improve the way they address issues and solve problems. This can be done by conducting an iterative inquiry process that balances problem-solving actions implemented in a collaborative context with data-driven analysis to understand underlying causes of a problem (Johnson, 1976).

3.1.2   Descriptive study

Descriptive research, also known as statistical research, describes data and characteristics about the population or phenomenon being studied. It answers the questions who, what, where, when and how. A descriptive study is undertaken to describe the characteristics of the variables of interest in a situation, for example, the mean and standard deviation of the sample under study as to the variable of interest. Quite frequently, descriptive studies are undertaken in organisations to learn about and describe the characteristics of a group of employees. Descriptive studies are also undertaken to understand the characteristics of organisations that follow certain common practices. The goal of a descriptive study, therefore, is to describe relevant aspects of the phenomena of interest from an individual, organisational, industry-oriented or other perspective (Sekaran, 2003).

Although the data description is factual, accurate and systematic, the research cannot describe what caused a situation. Thus, descriptive research cannot be used to create a causal relationship, where one variable affects another. The descriptive dimension in descriptive research is used for frequencies, averages and other statistical calculations. Often the best approach, before writing descriptive research, is to conduct a survey investigation. Qualitative research often has the aim of description and researchers may follow up with examinations of why the observations exist and what the implications of the findings are.

3.1.3   Correlational study

Correlational research attempts to determine whether and to what degree a relationship exists between two or more quantifiable (numerical) variables. Correlation implies prediction of the value of one variable if we know the value of the other correlated variable, but does not necessarily imply full causation. The reason why correlation does not necessarily imply full causality is that a third variable may be involved of which we are not aware. However, correlational research can imply partial correlation in terms of prediction of percentage of variation in variable B due to variable A.

Unlike in experimental research, in correlational research, researchers do not influence variables but only measure them and look for relations (correlations) between some sets of variables. In contrast, in experimental research, researchers manipulate some variables and then measure the effects of this manipulation on other variables. Correlation research explores the existence of a relationship in terms of direction and strength as follows (Cohen, 1988):

- A correlation has direction and can be either positive or negative. With a positive correlation, individuals who score high (or low) on one measure tend to score similarly on the other measure. With negative relationships, an individual who scores high on one measure tends to score low on the other (or vice versa).
- A correlation can differ in the degree or strength of the relationship. Zero indicates no relationship between the two measures. A correlation coefficient of 1.00 or –1.00 indicates a perfectly positive or negative relationship. The strength of a correlation can be anywhere between 0 and ±1.00.

Correlational research is often conducted at the beginning of a study and before exploratory research. It has three types: observational, survey and archival. These are described in Sections 3.1.3.1 to 3.1.3.3 followed by the objectives of correlational study and interpretation of correlations.

3.1.3.1   *Observational research*

Many types of studies could be defined as observational research, such as case studies. The primary characteristic of observational research is that phenomena are being observed and recorded. Observational research involves observing and recording the variables of interest in the natural environment without interference or manipulation by the experimenter. The advantages of observational research are

1. It gives the experimenter the opportunity to view the variable of interest in a natural setting.
2. It can offer ideas for further research.
3. It may be the only option if laboratory experimentation is not possible.

The disadvantages of the observational research are

1. It can be time consuming and expensive.
2. It does not allow for scientific control of variables.
3. Experimenters cannot control extraneous variables.
4. Subjects may be aware of the observer and may act differently as a result.

There are two types of observational research: differential and causal comparative research.

3.1.3.1.1   Differential research   Differential research is used when the manipulation of an independent variable is impractical. It observes two or more groups that are differentiated on the basis of some preexisting variable. Differential research has the following basic features:

1. Group differences existed before any research study was conducted.
2. It uses statistical techniques similar to correlation.

3.1.3.1.2   Causal comparative research   Causal-comparative research is similar to experimental research in that it attempts to establish cause-and-effect relationships among the independent and dependent variables, but in the case of causal-comparative research, the cause, that is, the independent variable, is not under the control of the experimenter. The experimenter has to take the values of the independent variable as they come. The dependent variable in a study is the outcome variable. This type of research usually involves group comparisons.

3.1.3.2   *Survey research*

Surveys and questionnaires are one of the most common methods used in research. In this method, a random sample of participants completes a survey, test or questionnaire that relates to the variables of interest. Random sampling is a vital part of ensuring the generalisability of the survey results. Survey research has the following advantages:

1. It is fast, inexpensive and easy; researchers can collect large amounts of data in a relatively short amount of time.
2. It is more flexible than some other methods.

It has the following disadvantages:

1. It can be affected by an unrepresentative sample or poor survey questions.
2. Participants can affect the outcome. Some participants try to please the researcher, give false responses to make themselves look better or have mistaken memories.

3.1.3.3   *Archival research*

Archival research is performed by analysing studies conducted by other researchers or by looking at historical records. The advantages of archival research are

1. Enormous amounts of data provide a better view of trends, relationships and outcomes.
2 It is often less expensive than other study methods.

The disadvantages of archival research are

1. Researchers have no control over how data are collected.
2. Previous research could be unreliable.

3.1.3.4   *Objectives of correlational studies*

Correlational studies help researchers to understand related events. They are usually conducted to (1) explore bivariate correlation, (2) explore regression, (3) explore inference for causality, (4) explore multiple correlation and (5) manipulate independent variables that are impractical to manipulate.

3.1.3.4.1   Exploring bivariate correlation   The bivariate correlation measures the relationship between two variables in terms of the strength and direction of this relationship. The degree of relationship, that is, how closely the variables are related, is usually expressed as a correlation coefficient. As the correlation coefficient moves toward either –1 or +1, the relationship gets stronger. A negative correlation means that as scores on one variable rise, scores on the other related variable decrease. A positive correlation indicates that the scores move together, both increasing or both decreasing.

3.1.3.4.2   Exploring regression   Regression analysis helps us to make predictions of how one variable might predict another. If there is a correlation between two variables, and we know the score on one variable, the second score can be predicted. Regression refers to how well we can make this prediction. As the correlation coefficients approach either –1 or +1, our predictions get better.

3.1.3.4.3   Exploring inference for causality   To make statements of cause and effect using correlational research methods, path analysis and cross-lagged panel designs are employed. Path analysis is used to determine which of a number of pathways connects one variable with another, and which path is the predominant path. Path analysis research uses correlational statistics to draw such conclusions. The cross-lagged panel design measures two variables at two points in time. It has been used, for example, to show that studying engineering science leads to conducting engineering research, more than the other way around.

There are two requirements to infer a causal relationship, as follows:

1. A statistically significant relationship exists between the variables.
2. The causal variable occurred before the other variable.

If there is no correlation, there is no causality, assuming the measures are valid and reliable. However, a proven a correlation between two variables does not necessarily mean that there is a causal relationship between them.

3.1.3.4.4   Exploring multiple correlation   This type of correlation extends regression and prediction analysis to include several more variables. The combination gives researchers more power to make accurate predictions. In this type of correlation, what researchers try to predict is called the criterion variable, and what they use to make the prediction, that is, the known variables, are called predictor variables.

3.1.3.4.5   Manipulating independent variables that are impractical to manipulate   The last possible aim of correlational studies is to manipulate independent variables that are impractical to manipulate for ethical or practical reasons. Correlational studies can be helpful to predict values of variables in such situations.

3.1.3.5 *Interpreting correlations*

In a project that includes several variables, beyond knowing the means and standard deviations of the dependent and independent variables, researchers often wish to know how one variable is related to another. The correlation coefficient (r) value ranges from –1 to 1. A correlation coefficient of –1 indicates a perfect negative relation between the variables under examination. If the correlation coefficient has a value of 0, there is no relation between the variables under examination. A correlation coefficient of 1 is interpreted as a perfect positive relation between the variables under examination. The square of the correlation coefficient (r^2) represents the percentage of variation in one of the two variables under investigation due to the other correlated variable. In an endeavour to improve correlational studies, partial correlation was introduced such that the influence of a potential third variable is removed when examining a correlation between two variables. Also, a multiple correlation is introduced to investigate a relationship between one variable and a set of variables.

3.1.3.6 *Advantages and limitations of correlational research*

The advantages and features of correlational research are as follows:

1. A consistent relationship can be used to predict variable's value or future events.
2. It provides data that are either consistent or inconsistent with some currently held scientific theory (correlation cannot prove a theory but can negate one).

The limitations of the correlational research are the following:

1. Variables must be quantifiable and usually represent at least an ordinal scale of measure.
2. Correlation does not necessarily mean causal inferences.

3.2 *Research methods and phases in engineering design management*

In engineering design management research, and because of its interdisciplinary nature, specific design methods are usually implemented. This section discusses research methods in engineering design management and research phases in engineering design management.

3.2.1   *Research methods in engineering design management*

There are three research methods that are usually employed in engineering design management research: quantitative, qualitative and a hybrid of the former two methods (Eckert et al., 2004).

3.2.1.1   *Quantitative research method*

The quantitative research method, also called scientific research, investigates quantitative properties and relationships. It employs measurement and counting to connect empirical observations to mathematical relationships and theories. This research method has the attributes of conciseness, objectivity and repeatability and leads to adherence to research designs.

3.2.1.2   *Qualitative research method*

Qualitative research is rather subjective. It is associated with openness and thoroughness in collecting data. This research method adheres to a philosophical perspective. The researcher becomes the primary instrument of investigation. Therefore, the process by which research is conducted should be carefully selected to minimise the effect of any researcher bias on its conclusion.

3.2.1.3   *Hybrid quantitative–qualitative research method*

The hybrid quantitative–qualitative research method incorporates elements of both quantitative and qualitative research methods. This has been exhibited by the interdisciplinary nature of the engineering design management research that sometimes necessitates different forms of knowledge in each research phase to address the research questions. Hybridising these two paradigms is a challenge common to much design research, which aims to both describe and improve the effectiveness of the design process.

3.2.2   *Research phases in engineering design management*

There are four research phases in the engineering design research methodology (Blessing et al., 1995; Eckert et al., 2003):

1. Research goals and criteria, in which the goals of the research are established and accordingly criteria are formulated by which the research results will be evaluated
2. Description I, in which a detailed understanding of the research gap and problem is formulated through literature review; this phase should result in an expanded problem definition and a description of the context within which the research conclusions may be considered valid.

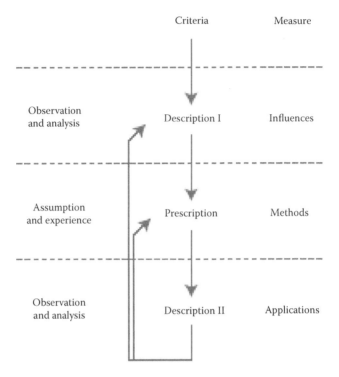

Figure 3.1 The four research phases in the engineering design research methodology. (From Blessing, L. T. M., Charkrabarti, A., Wallace, K. M. (1995). "A design research methodology." In *Proceedings of the 10th International Conference on Engineering Design (ICED'95)*, Prague, Czech Republic, Vol. 1, pp. 50–55.)

3. Prescription, in which methods that distill theories (which describe the nature of the problem) into pragmatic approaches or describe best practice are developed, or computer tools are developed
4. Description II, in which the research results and approach are validated against the research criteria; case studies are usually incorporated in this phase to validate the research results further.

Figure 3.1 illustrates these four research phases and shows their iterative nature.

3.3 Other classifications of research methodologies to understand the product development process

The research methodologies used to understand the product development process can also be classified into inductive research and deductive research.

3.3.1 Inductive research

Inductive research is the research process in which the premises, that is, the given statements, of a reasoning argument are believed to support the conclusion but they do not ensure the truth of the conclusion. Inductive research makes generalisations based on limited individual instances, such as number of observations of recurring phenomenal patterns. It tries to reach a conclusion about unobserved entities on the basis of those that have been observed. In other words, it tries to make general evidence on the basis of a particular truth. Therefore, the logic of inductive research is controversial because it often could be false (Manktelow, 1999).

3.3.2 Deductive research

Deductive research is more reliable and widely used than inductive research. It depends on its premises to reach a conclusion that must be true if the premises are true. In other words, deductive reasoning logic progresses from general evidence to a particular truth or conclusion. Deductive research stems from a large number of particular occurrences to a general rule (Sekaran, 2003).

3.4 Types of data

Research data that are collected can be classified into either qualitative or quantitative data, each with its own characteristics.

3.4.1 Qualitative data

Qualitative data refer to information gathered in a narrative form through interviews and/or observations. The qualitative approach involves collection of narrative data, for example, nonnumerical data, on many variables over a period of time to gain insights into phenomena of interest or data analysis. Examples of qualitative studies can include a case study of managerial involvement at an industrial company and a case study of researchers who excel despite nonfacilitating environments. In qualitative studies only small samples are invariably chosen in view of the indepth nature of the study. Therefore, generalisability of the findings is very restricted.

3.4.1.1 Operationalisation

Operationalisation means rendering a concept measurable. This is done by looking at the behavioural dimensions, facets or properties denoted by the concept. These are then translated into observable and measurable elements so as to develop an index of measurement of the concept.

3.4.2   Quantitative data

Quantitative approaches involve collection of numerical data to explain, predict and/or control phenomena of interest. The analysis of this type of data is mainly statistical. In addition, a quantitative study involves collecting data to test hypotheses or answer questions regarding subjects of study. In contrast with the qualitative approach, the data are numerical. The data are typically collected through a questionnaire, an interview or observation. Examples of quantitative studies can include how a design manager in an industrial company feels about a 12-month project.

3.5   Sampling

Because correlational, or quantitative research, is statistics based, effective sampling is key to reliable and valid correlational research. A sample consists of a subset of the population. A theoretical list (an actual list may not exist) of individuals or elements who make up a population is called a sampling frame. When conducting research, one often uses a sample of the population instead of the entire population. There are at least four major reasons to sample. First, it is usually too costly to test the entire population. Second, it may be impossible to test the entire population. Third, testing the entire population often produces error. If one individual rates the entire sample, there will be some measure of consistency because it is the same individual who rates the same sample. If many raters are used, as may be needed when testing the entire population, this introduces a source of error. The final reason to sample is that testing may be destructive. For instance, one probably would not want to buy a robot that had been crash tested. Sampling includes the following issues: probability and nonprobability sampling, sample size, sample representativeness and sampling problems (Sekaran, 2003).

3.5.1   Probability and nonprobability sampling design

There are two major types of sampling designs: probability and nonprobability sampling. In probability sampling, the elements in the population have some known or predetermined chance or probability of being selected as sample subjects. In nonprobability sampling, the elements do not have a known chance of being selected as subjects.

3.5.1.1   Probability sampling design

In probability sampling, the elements in the population have some known chance or probability of being selected as sample subjects. Probability sampling designs are used when the representativeness of the sample is of importance in the interests of wider generalisability. Probability sampling can be either unrestricted or restricted in nature.

3.5.1.1.1   Unrestricted probability sampling or simple random sampling   In unrestricted probability sampling the simple random sampling is used, in which every element in the population has a known and equal chance of being selected as a subject. If there are 1000 elements in the population, the first piece drawn will have a 1/1000 chance of being drawn, the next one a 1/999 chance of being drawn and so on. In the simple random sample method, all subjects or elements have an equal probability of being selected. There are two major ways of conducting a random sample. The first is to consult a random number table, and the second is to have the computer select a random sample.

3.5.1.1.2   Restricted or complex probability sampling   As an alternative to the simple random sampling design, several complex probability sampling, that is, restricted probability, designs can be used. The three most common complex probability sampling designs are systematic, stratified random and cluster sampling.

3.5.1.1.3   Systematic sampling   A systematic sample is conducted by randomly selecting a first case on a list of the population and then proceeding every Nth case until the sample is selected. This is particularly useful if the list of the population is long. For example, if your list was the designers' list of names, it would be easier to start at perhaps the 13th person, and then select every 40th person from that point on.

3.5.1.1.4   Stratified random sampling   Although sampling helps to estimate population parameters, there may be identifiable subgroups of elements within the population that may be expected to have different parameters on a variable of interest to the researcher. Stratified random sampling, as its name implies, involves a process of stratification or segregation, followed by random selection of subjects from each stratum, that is, each subpopulation. In a stratified sample, we sample either proportionately or equally to represent various strata or subpopulations. The population is first divided into mutually exclusive groups that are relevant, appropriate and meaningful in the context of the study. Stratification also helps when a research question such as the following is to be answered: Are machinists more accident prone than clerical workers? Stratification is an efficient research sampling design because it provides more information with a given sample size. Yet, conducting stratification can be costly. Once the population has been stratified in some meaningful way, a sample of members from each stratum can be drawn using either a simple random sampling or a systematic sampling procedure. The subjects drawn from each stratum can be either proportionate or disproportionate to the number of elements in the stratum. Disproportionate sampling is performed when some stratum or strata are too small or too large, or when there is more variability suspected within a particular stratum.

3.5.1.1.5 Cluster sampling When several groups with intragroup heterogeneity, that is, within each group, and intergroup homogeneity, that is, among groups, are found, then a random sampling of the clusters or groups ideally can be done and information gathered from each of the members in the randomly chosen clusters. In cluster sampling, a random sample of strata is taken and then every member of the group is surveyed. Cluster samples offer more heterogeneity within and more homogeneity among groups, that is, the reverse of what is found in the stratified random sampling where there is homogeneity within each group and heterogeneity across groups. However, cluster sampling is subject to greater biases and is the least generalisable of all the probability sampling designs, because most naturally occurring clusters in the organisational context do not contain heterogeneous elements.

3.5.1.2 Nonprobability sampling design
In nonprobability sampling, the elements do not have a known or predetermined chance of being selected as subjects. When the demands of time and resources and the purpose of the study rather than generalisability become critical, nonprobability sampling is generally used. Because there are no probabilities attached to the elements in the population, the findings from the study of the sample cannot be confidently generalised to the population. However, researchers may at times be less concerned about generalisability than obtaining some preliminary information in a quick and inexpensive way. They would then resort to nonprobability sampling. There are two types of nonprobability sampling: convenience and purposive sampling.

3.5.1.2.1 Convenience sampling As its name implies, convenience sampling refers to the collection of information from members of the population who are conveniently available to provide it. Convenience sampling is most often used during the exploratory phase of a research project and is perhaps the best way of getting some basic information quickly and efficiently. This is by far the most often used sample procedure. It is also by far the most biased sampling procedure as it is not random; that is, not everyone in the population has an equal chance of being selected to participate in the study.

3.5.1.2.2 Purposive sampling Instead of obtaining information from those who are most readily or conveniently available, it might sometimes become necessary to obtain information from specific target groups. The sampling here is confined to specific types of people who can provide the desired information, either because they are the only ones who have it or conform to some criteria set by the researcher. This type of sampling design is called purposive sampling, and the two major types are judgment and quota sampling. Judgment sampling involves the choice of subjects who are most advantageously placed or in the best position to provide the information required. Judgment sampling calls for special efforts to locate and gain access to the individuals who do have the required information. Quota sampling ensures that certain groups are adequately represented in the study through the assignment of a quota. Generally, the quota fixed for each subgroup is based on the total numbers of each group in the population. However, because this is a nonprobability sampling plan, the results are not generalisable to the population.

3.5.2 Sample size

A reliable and valid sample should enable us to generalise the findings from the sample to the population under investigation. The sample statistics should be reliable estimates and reflect the population parameters as closely as possible within a narrow margin of error. The investigators need not necessarily take several different samples to estimate this variability, as long as the sampling precision and confidence levels are appropriate. Precision is a function of the range of variability in the sampling distribution of the sample mean. Where precision denotes how close we estimate the population parameter based on the sample statistic, confidence denotes how certain we are that our estimates will really hold true for the population. There is a trade-off between precision and confidence for any given sample size. The level of confidence can range from 0 to 100%. A 95% confidence is the conventionally accepted level for most research, most commonly expressed by denoting the significance level as $p \leq 0.05$. In other words, one can say that at least 95 times out of 100 the estimate will reflect the true population characteristic.

Sample size is determined using the Krejcic and Morgan simplified sample size decision table (Krejcie and Morgan, 1970), in which an empirically proven sample size is suggested for every population size. Too large a sample size, for example, greater than 500, could become a problem because one would be prone to committing type II statistical errors, that is, the error of accepting a false hypothesis. In other words, with too large a sample size, even weak relationships, for example, a correlation of 0.1 between two variables, might reach significance levels as if these relationships were true of the population.

There are usually two reasons behind making sampling analysis: to estimate the population parameters and/or to test hypotheses about population values. In both of these two scenarios there are five aspects that should be considered while making decisions on the sample size needed to do the research:

1. What is the population size?
2. How much precision is really needed in estimating the population characteristics of interest, that is, what is the margin of allowable error?
3. How much confidence is really needed, that is, how much chance can we take of making errors in estimating the population parameters?
4. To what extent is there variability in the population on the characteristics investigated?
5. What is the cost–benefit analysis for increasing the sample size?

3.5.3   Sample representativeness

Rarely will the sample be the exact replica of the population from which it is drawn. However, if the sample is chosen in a scientific way, researchers can be reasonably sure that the sample statistic is fairly close to the population parameter. Both sampling design and the sample size are important to establish the representativeness of the sample for generalisability. If the appropriate sampling design is not made, a large sample size will not, in itself, allow the findings to be generalised to the population. Likewise, unless the sample size is adequate for the desired level of precision and confidence, no sampling design, however sophisticated, can be useful to the researcher in meeting the objectives of the study. Hence, sampling decisions should consider both the sampling design and the sample size. There is often a trade-off between time and cost efficiencies (as achieved in nonprobability sampling designs) and precision efficiencies (as achieved in many probability sampling plans). The choice of a sampling plan thus depends on the objectives of the research, as well as on the extent and nature of efficiency desired.

3.5.4   Sampling problems

There are several potential sampling problems. First, there may be missing elements, that is, individuals who should be on the list but for some reason are not. For example, if the population consists of all individuals living in a particular city and the researcher uses the phone directory as the sampling frame or list, the researcher will miss individuals with unlisted numbers or those who cannot afford a phone. Foreign elements make up the second sampling problem. Elements that should not be included in the population and sample appear on the sampling list. Thus, if the researcher were to use property records to create a list of individuals living within a particular city, landlords who live elsewhere would be foreign elements. The third sampling problem is duplicates – elements that appear more than once on the sampling frame.

3.6 Statistical errors

In scientific research there are two different sorts of error:

1. Statistical error, which represents the difference between a computed, estimated or measured value and the true, or theoretically correct value that is caused by *random* and inherently unpredictable fluctuations in the measurement apparatus or in the system being studied.
2. Systematic error, which represents the difference between a computed, estimated or measured value and the true, or theoretically correct value that is caused by *nonrandom* fluctuations in the measuring apparatus or in the system being studied. Systematic error, once identified, can usually be eliminated. It can occur from an unknown source, as a sort of uncertainty.

The possible statistical errors in the decision process are either type I error, also referred to as α error, or type II error, also referred to as β error. These two sources of error are apparent in testing hypotheses. Hypothesis testing is the art of testing whether a variation between two sample distributions can be explained by chance or not. In testing hypotheses, the tested hypothesis is set as an alternative hypothesis to a null hypothesis that is used to falsify the alternative hypothesis. The null hypothesis corresponds to a presumed default, that is, state of nature. The alternative hypothesis corresponds to the opposite situation. The goal is to determine whether the null hypothesis can be discarded in favor of the alternative. These two sources of error in the statistical decision process, described in the following list, are based on whether or not a particular sample that was drawn from a certain population represents that population.

1. Type I, also known as an α error, is the the error of rejecting a null hypothesis when it is actually true. It occurs when we are observing a difference when in truth there is none.
2. Type II, also known as a β error, is the error of failing to reject a null hypothesis when the alternative hypothesis is the true state of nature. In other words, this is the error of failing to observe a difference when in truth there is one.

Therefore the hypothesis test must be able to (1) reduce the chance of rejecting a true hypothesis to as low a value as desired and (2) reject the tested hypothesis tested when it is likely to be false.

Usually type I error is more delicate than type II error. Hence, care is usually focussed on minimising the occurrence of type I error. Suppose the probability for a type I error is 1% or 5%; then there is a 1% or 5% chance that the observed variation is not true. This is called the *level of significance*. Although 1% or 5% might be an acceptable level of significance for one application, a different application can require a very different level. For example, the standard goal of six sigma is to achieve exactness by 4.5 standard deviations above or below the mean. That is, for a normally distributed process only 3.4 parts per million are allowed to be deficient. Any statistical result has a p-value that is compared with the significance level. In more common parlance, a type I error could be interpreted as a warning of falseness whereas a type II error could be interpreted as inadequate sensitivity.

3.6.1   Statistical power test

The power of a statistical test is the probability that the test will reject a false null hypothesis, that is, it will not make a type II error. As power increases, the chances of a type II error decrease. The probability of a type II error is referred to as (β). Therefore power is equal to $1 - \beta$. Power analysis can be done either before or after data are collected. Power analysis is conducted before the research study when the aim is to determine an appropriate sample size to achieve adequate power. Power analysis is also conducted after a study has been completed, and uses the obtained sample size and effect size, when the aim is to determine what statistical power was in the study, assuming the effect size in the sample is equal to the effect size in the population.

Despite the use of random samples, which will tend to mirror the population through mathematical properties such as the central limit theorem, there is always a chance that the samples will appear to support or refute a tested hypothesis when the reality is the opposite. This risk is quantified as the power of the test and as the statistical significance level used for the test. Statistical power depends on three aspects: (1) the statistical significance, (2) the size of the population and (3) the reliability of the data.

3.7   Data collection methods

Data can be collected using three major methods: (1) interviews, which include face-to-face and telephone interviews, as well as interviews through electronic media; (2) questionnaires, which are either personally administered, sent through the mail, or electronically administered and (3) observation of individuals and events with or without videotaping or audio recording.

3.7.1   Interviews

Interviewing respondents is intended primarily to obtain information on issues of interest. Interviews could be unstructured or structured. Unstructured interviews are so called because the interviewer does not enter the interview setting with a planned sequence of questions to be asked of the respondent. The objective of the unstructured interview is to bring some preliminary issues to the surface so that the researcher can determine what variables need further in-depth investigation. After conducting a sufficient number of such unstructured interviews with respondents at several levels and studying the data obtained, the researcher would know the variables that need greater focus and call for more in-depth information. This sets the stage for the interviewer to conduct further structured interviews, for which the variables would have been identified.

Structured interviews are those conducted when it is known at the outset what information is needed. The interviewer has a list of predetermined questions to be asked of the respondents either personally, through the telephone or through the electronic media. The questions are likely to focus on factors that had surfaced during the unstructured interviews and are considered relevant to the problem. As the respondents express their views, the researcher would note them down. The same questions will be asked of all respondents in the same manner.

The information obtained during the interviews should be as free as possible of bias. Bias refers to errors or inaccuracies in the data collected. Biases could be introduced by the interviewer, the interviewee or the situation. The interviewer could bias the data when the responses are misinterpreted or unintentionally encourages or discourages certain types of responses through gestures and facial expressions. Interviewees can bias the data when they do not express their true opinions. Also, if they do not understand the questions, they may hesitate to seek clarification. Biases could be situational as well, in terms of (1) nonparticipants, (2) trust levels and rapport established and (3) the physical setting of the interview. The main advantage of interviews is that the researcher can adapt the questions as necessary, clarify doubts and ensure that the responses are properly understood, by repeating or rephrasing the questions.

3.7.2   Questionnaires

A questionnaire is a preformulated written set of questions to which respondents record their answers, usually within rather closely defined alternatives. Questionnaires are an efficient data collection mechanism when the researcher knows exactly what is required and how to measure the variables of interest. Questionnaires can be administered personally, mailed to the respondents or distributed electronically.

When the survey is confined to a local area, and the organisation is willing and able to assemble groups of respondents to respond to the questionnaires at the workplace, a good way to collect data is to administer the questionnaires personally. The main advantage of this is that the researcher or a member of the research team can collect all the completed responses within a short period of time. The researcher is also afforded the opportunity to motivate the respondents to offer frank answers. Administering questionnaires to large numbers of individuals at the same time is less expensive and consumes less time than interviewing; it also does not require as much skill to administer the questionnaire as to conduct interviews. Questionnaires are most useful as a data collection method especially when large numbers of people are to be reached in different geographical regions. They are a popular method of collecting data because researchers can obtain information fairly and easily and the questionnaire responses are easily coded.

The main advantage of mail questionnaires is that a wide geographical area can be covered in the survey. Respondents can complete them at their convenience and at their own pace. However, the return rate of mail questionnaires is typically low. Another disadvantage of the mail questionnaire is that any doubts the respondents might have cannot be clarified. Also, with the very low return rates it is difficult to establish the representativeness of the sample. Electronically distributed questionnaires are similar to mail questionnaires and have the following additional advantages:

1. They are faster than mail questionnaires.
2. They are less costly than mail questionnaires.
3. Any doubts the respondents might have can be clarified.

Sound questionnaire design principles should focus on three areas: (1) the wording of the questions; (2) how the variables will be categorised, scaled and coded after receipt of the responses and (3) the general appearance of the questionnaire. These three issues are important to minimise biases in research. The principles of wording in questionnaires refer to the following five factors: (1) the content and wording of the questionnaire, (2) the type of questions asked, (3) the form of questions asked, (4) the sequencing of the questions and (5) the personal data sought from the respondents.

3.7.2.1   *Content and wording of the questionnaire*

The language of the questionnaire should approximate the level of understanding of the respondents. Some terminologies and jargon could not be understood; thus it is important to word the questions in a way that could be understood by the respondent.

3.7.2.2 *Types of questions asked*

Type of question refers to whether the question will be open-ended or closed-ended. Open-ended questions allow respondents to answer them in any way they choose. An example of an open-ended question is asking the respondent to state five things that are interesting and challenging in his or her job. A closed question, in contrast, would ask the respondents to make choices among a set of alternatives given by the researcher. Closed questions help the respondents to make quick decisions to choose among the several alternatives before them. They also help the researcher to code the information easily for subsequent analysis. Care has to be taken to ensure that the alternatives are mutually exclusive and collectively exhaustive; that is, there can be overlapping categories in the alternatives, or not all possible alternatives are given.

3.7.2.3 *Form of questions asked*

Form of questions refers to positively and negatively worded questions. Instead of phrasing all questions positively, it is advisable to include some negatively worded questions as well. Having both positively and negatively worded questions, the respondent could resist any likely tendency to mechanically circle the points towards one end of the scale. In case this does still happen, the researcher has an opportunity to detect such biases. A question that lends itself to different possible responses to its subparts is called a double-barrelled question. Such questions should be avoided and two or more separate questions asked instead. Questions should be clearly stated and not worded ambiguously. Also, they should not require respondents to recall experiences from the past that are hazy in their memory, as answers to such questions might contain bias.

Questions should not be phrased in such a way that they lead the respondents to give the responses that the researcher would like or want them to give, for example, eliciting socially desirable responses. Another type of bias occurs when questions are phrased in an emotionally charged manner. If the purpose of the question is twofold, then these are the two specific questions that need to be asked. Finally, simple, short questions are preferable to long ones.

3.7.2.4 *Sequencing of the questions*

The sequence of questions in the questionnaire should be such that the respondent is led from questions of a general nature to those that are more specific, and from questions that are relatively easy to answer to those that are progressively more difficult. This funnel approach, as it is called (Festinger and Katz, 1996), facilitates the easy and smooth progress of the respondent through the items in the questionnaire. The progression from general to specific questions might mean that the respondent is first asked questions of a global nature that pertain to the organisation, and then is asked more incisive questions regarding the specific job and the like.

3.7.2.5   *Personal data sought from the respondents*

Classification data, also known as personal information or demographic questions, elicit information such as age. Unless absolutely necessary, it is best not to ask for the name of the respondent. If, however, the questionnaire has to be identified with the respondents for any reason, then the questionnaire could be numbered and connected by the researcher to the respondent's name, in a separately maintained and private document. This procedure should be clearly explained to the respondent. The reason for using the numerical system in questionnaires is to ensure the anonymity of the respondent, should the questionnaires fall into the hands of someone in the organisation.

The general appearance of the questionnaire is an important design principle. A neat questionnaire with appropriate introduction, instructions and well-arrayed set of questions and response alternatives will make it easier for the respondents to answer them.

3.7.3   *Observational surveys*

Whereas interviews and questionnaires elicit responses from the subjects, it is possible to gather data without asking questions of respondents. People can be observed in their natural work environment or in the laboratory setting, and their activities and behaviours or other items of interest can be noted and recorded. Observational studies can be structured or unstructured. All phenomena of interest are systematically recorded and quality control can be exercised by eliminating biases. Observational studies can provide rich data and offer understanding of phenomena observed.

The following are among the advantages of observational studies:

1. The data obtained through observation of events as they normally occur are generally more reliable and free from respondent bias.
2. In observational studies, it is easier to note the effects of environmental influence on specific outcomes.
3. It is easier to observe certain groups of individuals.

The following are among the disadvantages of observational studies:

1. It is necessary for the observer to be physically present, often for prolonged periods of time.
2. This method of collecting data is not only slow, but also tedious and expensive.

3. Because of the long periods for which subjects are observed, observer fatigue could easily set in, which might bias the recorded data.
4. Though moods, feelings and attitudes can be guessed by observing facial expressions and other nonverbal behaviours, the cognitive thought processes of individuals cannot be captured.

chapter four

Engineering design experiments

Quasi-experiments are adopted in this book to collect empirical data to verify the research hypotheses and answer the research questions. This chapter presents the exploratory quasi-experiments that were conducted in this research area and their outcomes.

4.1 Selected industrial sectors for the experiments

To explore the practicality of the present work, two industrial sectors are selected in which product attributes are investigated. Because the two major streams of industrial projects are labour-intensive projects and capital-intensive projects, both are represented in this book (Wegner, 1984). The structural design and construction engineering industrial sector and projects are selected to represent the labour-intensive industrial projects. The robotics and flexible manufacturing systems industrial sector and projects are selected to represent the capital-intensive industrial projects.

4.1.1 Investigated product attributes in the structural design and construction engineering industrial sector

Market success has always been the ultimate goal of every project, and a construction project is no exception. The criteria used to evaluate the performance of a structural design and construction project have long been (1) time, (2) cost and (3) quality (Chan et al., 2002). Quality in this context means meeting the functional requirements, such as

1. Each part or structure carries a factor of safety that conforms to the relevant international standards.
2. Product volume is within the allowed space and lightweight (Cambridge University, Engineering Department [CUED], 2007a).

4.1.2 Investigated product attributes in the robotics and flexible manufacturing systems industrial sector

Although robotics and flexible manufacturing systems represent a capital-intensive industrial sector, they provide significant advantages to their customers. These advantages range from reducing operating costs, improving product quality and consistency and enhancing the quality of work for employees, to increasing production output rates, product manufacturing flexibility and reducing material waste and increasing yield (ABB Robotics Product Guide, 2008). In the robotics and flexible manufacturing systems industrial sector the product attributes become more complex than those of the structural design and construction engineering industrial sector. The current trend in robotics is to provide autonomous guided robots (Christensen et al., 2008), the quality of which is measured against the following attributes:

1. Minimum floor space requirements for agile motion
2. Adaptability to the surroundings and capability in making decisions accordingly, such as in case of path irregularity, missing a junction and encountering obstacles
3. Capability in recognising the position of an object
4. Capability in controlling the force used to grip an object
5. Flexibility for picking and depositing loads to a variety of station types and elevations
6. Capability in following a nonstraightforward path
7. Fast response
8. Stability
9. Accuracy
10. Payload capacity
11. Reliability
12. Maintainability
13. Safety (ABB Robotics Product Guide, 2008; Fanuc Robotics, 2008)

4.2   Research questions

As the literature review indicates, no other research empirically investigated the identified research gap. Therefore, the research goal is to address this gap through answering the following research questions, which are categorised into a primary research question and secondary research questions.

4.2.1   Primary research question

The primary question this book answers is as follows: Is there any correlation and causal relationship between the complex product design and development activities and strategies on the one hand, and the complex product success in terms of product attributes on the other?

4.2.2 Secondary research questions

The secondary questions raised in this context are

1. What are the product attributes that should be investigated?
2. What are the product design and development activities and strategies that have an impact on the product attributes?
3. What are the relationships between product attributes on the one hand and product design and development activities and strategies on the other?

4.3 Research hypotheses

This section presents the research hypotheses that are extracted from the research review and would be tested through research questionnaires. It is noteworthy in this context that the three subsystems are as follows:

- Mechanical
- Electronics
- Software

Also, for clarification purposes the technical term 'modular design' means splitting the overall system into modules that interface with each other to speed up the design development process, facilitate error-tracing and maintainability and minimise the impact of design iterations. Consequently, the meaning of 'system's modules' can be better conceived through the following four examples:

- Power train module
- Logic gates module
- Code implementation of line following functions
- Code implementation of class and methods of pickup and unloading actions

In addition, the following abbreviations are used in the rest of this book:

SD: system design dimension
PM: project management dimension
M: mechanical subsystem design dimension
E: electronics subsystem design dimension
S: software subsystem design dimension

The primary research hypothesis that is extracted from the research review and would be tested through research questionnaires is as follows:

Ho: There is no correlation between the product design and development activities and strategies on the one hand and the product success in terms of product technical attributes on the other.

Ha: There is a correlation and a causal relationship between the product design and development activities and strategies on the one hand and the product success in terms of product technical attributes on the other.

Table 4.1 presents the sub-hypotheses of the quasi-experiments. Wherever the statement of 'In the integrated design system' is found in this book, this implies that point is relevant only to the integrated design project (IDP) rather than to the structural design project (SDP).

4.4   Research validity and verification

Research validity and verification are important processes intended to ascertain whether research requirements are met. Research validity answers the question, 'Did we measure the right thing?' Research verification answers the question, 'Did we measure the thing right?' They are discussed in Sections 4.4.1 and 4.4.2.

4.4.1   Research verification and reliability

Research verification answers the following question: 'Did I measure the construct correctly?' The internal consistency of the items to be measured is indicative of verification of measuring the construct. In statistics, reliability is the consistency of a set of measurements and is often used to describe a test. In other words, reliability is the extent to which the measurements of a test remain consistent over repeated tests of the same subject under identical conditions. An experiment is reliable if it yields consistent results of the same measure. It is unreliable if repeated measurements give different results. It can also be interpreted as the lack of random error in measurement.

The homogeneity of the responses to the items in the measure indicates that the respondents attach the same overall meaning to each of these items. This can be seen by examining if the items and the subsets of items in the measuring instrument are highly correlated. This can be tested through the interitem consistency reliability test. To the degree that items are independent measures of the same concept, they will correlate with one another. The most popular test of interitem consistency reliability is the Cronbach's coefficient α. The closer the coefficient is to 1, the

Table 4.1 Sub-hypotheses of quasi-experiments

Research hypotheses	Literature reference	Subsystem type	Design phase	Design strategy/ activity	Operationalisation and mapping of questionnaire's questions
	System design dimension				
1. In the integrated design system, adopting a modular design, that is, splitting the overall system into modules that interface with each other to speed up the design development process, facilitates error-tracing and maintainability, minimises the impact of design iterations and improves the opportunity for achieving a better design, if design iterations are needed.	Gershenson and Prasad (1997); Smith and Reinertsen (1997); Clark and Baldwin (2000)	SD	Conceptual and preliminary design phases	Strategy	Question 6
2. In the integrated design system, adopting a systematic design approach, that is, (1) clarification of the design problem; (2) identifying requirements specification and design constraints; (3) sorting the requirements specification in a ranked order;	Pugh (1991a,b); Pahl and Beitz (1998); Sivaloganathan et al. (2000)	SD	Scoping and conceptual design phases	Activity	Question 7

(Continued)

Table 4.1 (Continued) Sub-hypotheses of quasi-experiments

Research hypotheses	Literature reference	Subsystem type	Design phase	Design strategy/ activity	Operationalisation and mapping of questionnaire's questions
(4) identifying functional requirements, for example, using an axiomatic design methodology, and identifying core design values; (5) constructing a table of design options based on brain storming; (6) making concept evaluation, for example, based on a weighted criteria evaluation matrix and (7) making an overall strategic decision of the conceptual design increases the likelihood of achieving a better design.					
3. In the integrated design system and in case of adopting the systematic design approach, paying special attention to the quality of performing each of the steps of the systematic design approach increases the likelihood of achieving a better design.	Sivaloganathan et al. (2000)	SD	Scoping and conceptual design phases	Activity	Question 8

(Continued)

Table 4.1 (Continued) Sub-hypotheses of quasi-experiments

Research hypotheses	Literature reference	Subsystem type	Design phase	Design strategy/ activity	Operationalisation and mapping of questionnaire's questions
4. In the integrated design system, setting an error-proof operational design strategy, for example, constructing a 3D computer-aided design (CAD) model of the system assembly drawing minimises waste of time and resources caused by avoidable mistakes and improves the opportunity for achievement of a better design.	Chao et al. (2005)	SD	Preliminary and detailed design phases	Strategy	Question 9
5. In the integrated design system, designers aim at striking a balance between maximising functionality of the product final concept on one hand and minimising waste of time and resources by minimising design iterations on the other hand.	Eppinger (2001)	SD	Conceptual and preliminary design phases	Strategy	Question 11

(Continued)

Table 4.1 (Continued) Sub-hypotheses of quasi-experiments

Research hypotheses	Literature reference	Subsystem type	Design phase	Design strategy/activity	Operationalisation and mapping of questionnaire's questions
6. In the integrated design system, adopting top-down system structure decomposition to analyse the functional structure of the product and then map it to the requirements specification and consequently come up with the functional requirements of the product improves the opportunity for achieving a better design.	Bernard (1999)	SD	Scoping phase	Activity	Question 12
7. In the integrated design system, the design team aims at striking a balance between fast response on one hand and stability, accuracy and payload fulfilment of the final concept of the product on the other.	Yavuz (2007); Fanuc Robotics (2008)	SD	Conceptual and preliminary design phases	Strategy	Question 13
8. In the integrated design system, the design team considers reliability of the final product in the design process, in terms of the ability of the product to perform its required functions under stated conditions for a specified product service time.	Coulibaly et al. (2008); Prabhakar Murthy et al. (2008)	SD	Conceptual and preliminary design phases	Strategy	Question 14

(Continued)

Table 4.1 (Continued) Sub-hypotheses of quasi-experiments

Research hypotheses	Literature reference	Subsystem type	Design phase	Design strategy/ activity	Operationalisation and mapping of questionnaire's questions
9. In the integrated design system, the design team considers maintainability of the final product in the design process, in terms of the ease with which maintenance of the functional units of the product can be performed.	Coulibaly et al. (2008)	SD	Conceptual and preliminary design phases	Strategy	Question 15
10. In the integrated design system, the design team considers safety of use of the final product in the design process, in terms of accident prevention, risk identification and risk control and/or minimisation.	Coulibaly et al. (2008)	SD	Conceptual and preliminary design phases	Strategy	Question 16
11. In the integrated design system, those who predict working scenarios of risk, for example, inappropriate lighting conditions, or missing a junction on the route, and accordingly prepared action plans, for example, error-recovery tactic of tolerant triggering threshold, or memorising last state of all input signals and consequently correcting the path based on the expected	Rostami et al. (2005)	SD	Preliminary and detailed design phases	Strategy	Question 17

(Continued)

Table 4.1 (Continued) Sub-hypotheses of quasi-experiments

Research hypotheses	Literature reference	Subsystem type	Design phase	Design strategy/ activity	Operationalisation and mapping of questionnaire's questions
milestone on the route plan, improve performance robustness of the product in its final concept and obtain higher performance scores.					
Project management dimension					
12. Visiting, that is, looking at, and reviewing similar designs that solved similar design problems, improves the opportunity for achieving a better design.	Dahl et al. (2001)	PM	Scoping phase	Activity	Question 22
13. Adopting cross fertilisation of design concepts among the design team, such as sharing ideas and/or modifying each other's ideas, increases the likelihood of achieving a better design.	Amon et al. (1995)	PM	Conceptual and preliminary design phases	Activity	Question 23

(Continued)

Table 4.1 (Continued) Sub-hypotheses of quasi-experiments

Research hypotheses	Literature reference	Subsystem type	Design phase	Design strategy/ activity	Operationalisation and mapping of questionnaire's questions
14. In the integrated design system, having a multidisciplinary team of designers, each of whom is aware of more than one of the relevant disciplines such as material science, design development approaches, leadership skills and so forth, and is aware of the overlap and intersections between them, improves the opportunity for achieving a better design and minimising the risk of mistakes.	Amon et al. (1995)	PM	Scoping phase	Activity	Question 25
15. Having provisions for unforeseen problems to minimise vulnerability of the design development process to the influence of external factors improves the opportunity for achieving a better design.	Browning et al. (2005)	PM	Scoping phase	Strategy	Question 26

(Continued)

Table 4.1 (Continued) Sub-hypotheses of quasi-experiments

Research hypotheses	Literature reference	Subsystem type	Design phase	Design strategy/activity	Operationalisation and mapping of questionnaire's questions
16. In the integrated design system, setting an operational design strategy of modular testing, that is, testing the deliverables between the system submodules as a way of verifying conformity of these system submodules to the conceptual functional requirements to detect mistakes as early as possible and to minimise the impact of mistakes on the successful completion of the design project improves the opportunity for achieving a better design.	Lévárdy and Browning (2005)	PM	Conceptual and preliminary design phases	Strategy	Question 27
17. In the integrated design system, adopting testable design interdeliverables within and among system modules based on project milestones, to detect mistakes as early as possible and to minimise their impact on the successful completion of the design project, improves the opportunity for achieving a better design.	Huang (2000); Lévárdy and Browning (2005)	PM	Preliminary and detailed design phases	Activity	Question 29

(Continued)

Table 4.1 (Continued) Sub-hypotheses of quasi-experiments

Research hypotheses	Literature reference	Subsystem type	Design phase	Design strategy/ activity	Operationalisation and mapping of questionnaire's questions
18. In the integrated design system, adopting live meetings of the design team in the conceptual and preliminary design phases and e-mails in the detailed design phase as the principal communication methods, to strike a balance between the synergy of face-to-face meetings on one hand and the time effectiveness and message delivery effectiveness of the electronic means of communication, such as e-mails, on the other, improves the opportunity for achieving a better design.	Court (1998)	PM	Conceptual, preliminary and detailed design phases	Activity	Question 30
19. In the integrated design system, adopting a decentralised decision-making strategy, by empowering and authorising subteams to make tactical decisions without the need to refer them to the team leader, rather than strategic decisions that can degrade performance in this case, improves the agility of the design process.	Krishnan (1998); Pearce (1999)	PM	Conceptual, preliminary and detailed design phases	Strategy	Question 31

(Continued)

Table 4.1 (Continued) Sub-hypotheses of quasi-experiments

Research hypotheses	Literature reference	Subsystem type	Design phase	Design strategy/ activity	Operationalisation and mapping of questionnaire's questions
20. Adopting both sequential and concurrent design activities according to the nature of each task and the preceding relationships between them, to maximise utilisation of resources and shorten the design development time, improves the opportunity of achieving a better design.	Browning and Eppinger (2002)	PM	Scoping phase	Strategy	Question 32
21. In the integrated design system, the design team is more interested in meeting project delivery deadlines than in achieving the project objectives to a specified cost.	Mahmoud-Jouini et al. (2004); Clarkson and Eckert (2005)	PM	Scoping, conceptual, preliminary and detailed design phases	Strategy	Question 33
22. In the integrated design system, the design team adopts redeployment of the human resources of the design team to cope with fluctuations in the workload of project activities.	Browning and Eppinger (2002); Mahmoud-Jouini et al. (2004)	PM	Conceptual, preliminary and detailed design phases	Strategy	Question 35

(Continued)

Table 4.1 (Continued) Sub-hypotheses of quasi-experiments

Research hypotheses	Literature reference	Subsystem type	Design phase	Design strategy/ activity	Operationalisation and mapping of questionnaire's questions
23. In the integrated design system, given the intertwined and overlapping nature of the subsystems of the product, for example, the mechanical-electronic interconnection for robot speed, the design team adopts modular deliverables and testing, that is, testing deliverables of each module, rather than subsystem deliverables and testing.	Clark and Baldwin (2000)	PM	Preliminary and detailed design phases	Activity	Question 36
24. In the integrated design system, the design team conducts a design research activity (e.g. research literature review, review of previous literature review, review of previous reports, etc.) and/or Internet literature review search to become acquainted with what other designers have done to design and implement similar products and to become acquainted with the state of the art of the design of the required product.	Dahl et al. (2001)	PM	Scoping phase	Activity	Question 38

(Continued)

Table 4.1 (Continued) Sub-hypotheses of quasi-experiments

Research hypotheses	Literature reference	Subsystem type	Design phase	Design strategy/ activity	Operationalisation and mapping of questionnaire's questions
25. In the integrated design system, total time duration of the project is the most important constraint, rather than a resource, because designers cannot control the rate at which it is expended and imposed on the project. Those who considered the total time duration of the project as the most important constraint and consequently were more interested in the critical path, the longest sequence of time-constrained design tasks in the project plan, rather than in the critical chain, the longest sequence of resource-constrained design tasks in the project plan, received a higher performance score.	Leach (2000)	PM	Scoping phase	Strategy	Question 40

(Continued)

Table 4.1 (Continued) Sub-hypotheses of quasi-experiments

Research hypotheses	Literature reference	Subsystem type	Design phase	Design strategy/ activity	Operationalisation and mapping of questionnaire's questions
26. In the integrated design system, those designers who did not document the outcome of their design discussions and consequently did not analyse that outcome lost a valuable source of design concept improvement and attained a less effective design concept and received a lower score on product performance.	Pahl and Beitz (1998)	PM	Conceptual and preliminary design phases	Activity	Question 41
27. In the integrated design system, the greater the number of experienced designers in the design team, the better they conceive the assumptions and constraints accompanied by similar designs.	Court (1998)	PM	Scoping phase	Strategy	Question 42

(Continued)

Table 4.1 (Continued) Sub-hypotheses of quasi-experiments

Research hypotheses	Literature reference	Subsystem type	Design phase	Design strategy/ activity	Operationalisation and mapping of questionnaire's questions
28. In the integrated design system, the greater the number of experienced designers in the design team, the more they prevent avoidable mistakes in the design process, such as mounting the more heat-sensitive electronic components first on the printed circuit board (PCB) prototype, and the more their design process becomes foolproof.	Court (1998)	PM	Scoping phase	Strategy	Question 43
29. In the integrated design system, the more the design team meets up collectively in the conceptual design phase than in detailed design phase, the greater is the likelihood of achieving a more effective design concept.	Court (1998)	PM	Conceptual and detailed design phases	Activity	Question 44

(Continued)

Table 4.1 (Continued) Sub-hypotheses of quasi-experiments

Research hypotheses	Literature reference	Subsystem type	Design phase	Design strategy/activity	Operationalisation and mapping of questionnaire's questions
Mechanical subsystem design dimension					
30. Considering manufacturability starting from the conceptual design phase onwards in the design process, to minimise waste of time and resources and to eliminate relevant avoidable mistakes and unnecessary iterations, improves the opportunity for achieving a better design.	Pahl and Beitz (1998)	M	Conceptual, preliminary and detailed design phases	Strategy	Question 53
31. Checking accuracy of manufacturing and assembly of the final prototype, to avoid unexpected failure due to manufacturing defects and/or assembly mistakes, improves the opportunity of achieving a better design.	Pahl and Beitz (1998)	M	Detailed design phase	Activity	Question 54
Electronics subsystem design dimension					
32. In the integrated design system, adopting electronic distinguishing systems is more reliable than adopting mechanical distinguishing systems.	Ascher (2007)	E	Conceptual, preliminary and detailed design phases	Strategy	Question 4

(Continued)

Table 4.1 (Continued) Sub-hypotheses of quasi-experiments

Research hypotheses	Literature reference	Subsystem type	Design phase	Design strategy/ activity	Operationalisation and mapping of questionnaire's questions
Software subsystem design dimension					
33. In the integrated design system, adopting quick testing of interdeliverables between the modules of the software draft code, and extensive testing of the overall software draft code on a prototype PCB or on an equivalent facility, to strike a balance between minimising cost of test and detecting mistakes as early as possible, improves the opportunity for achieving a better design.	Lévárdy and Browning (2005)	S	Preliminary and detailed design phases	Activity	Question 60
34. In the integrated design system, software designers prefer object-oriented programming code implementation to structured programming code implementation because of the effectiveness of the former to achieve a better design because functions and subroutines are	Rob (2004)	S	Preliminary and detailed design phases	Strategy	Question 64

(Continued)

Table 4.1 (Continued) Sub-hypotheses of quasi-experiments

Research hypotheses	Literature reference	Subsystem type	Design phase	Design strategy/activity	Operationalisation and mapping of questionnaire's questions
less effective as to reusability, scalability and manageability. In addition, object-oriented programming code implementation offers the advantage of modularisation of classes.					
35. In the integrated design system, software designers aim at implementing explanatory annotation of the software code for purposes of software code debugging and maintenance.	Wasserman et al. (1990)	S	Preliminary and detailed design phases	Activity	Question 65
36. In the integrated design system, adopting a parameterisation design strategy, that is, changing the value of one parameter, changes the values of many related parameters accordingly, to facilitate software code debugging, maintainability and/or scalability of the product final concept, improves the opportunity for achieving a better design.	Loughran and Rashid (2004)	S	Preliminary and detailed design phases	Strategy	Question 66

better is the internal consistency reliability. An α value less than 0.7 indicates that the reliability of the data is low. An α value between 0.7 and 0.8 indicates that the reliability of the data is moderate. An α value larger than 0.8 indicates that the reliability of the data is high.

4.4.2   Research validity

A key concept relevant to the research methodology is the validity of research, which refers to the degree to which we are confident in the results that they measured the right thing that was theoretically intended meaningfully, reasonably and invariably with sample size. There are four aspects of research validity: (1) statistical conclusion, (2) internal, (3) construct and (4) external validity (Cook and Campbell, 1979b). Each of these is briefly defined and described in Sections 4.4.2.1 to 4.4.2.4.

4.4.2.1   Statistical conclusion validity

Statistical conclusion validity refers to whether the resulting relationships are meaningful and reasonable. Essentially, the question that is being asked is, 'Does variable A covary with variable B?' If a study has good statistical conclusion validity, the answer that the results provide to such a question should be meaningful and reasonable. An example of issues or problems that would threaten statistical conclusion validity is a small sample size, as it is more difficult to find meaningful relationships with a small number of subjects.

4.4.2.2   Internal validity

Internal validity refers to the confidence we place in the cause-and-effect relationship. If a study lacks internal validity, one cannot make cause-and-effect statements based on the results because they would be descriptive but not causal. Essentially, the question that is being asked is, 'To what extent does the research design permit us to say that the independent variable A causes a change in the dependent variable B?' There are many potential threats to internal validity. For example, if the research subjects, for example, sample members, are not randomly selected, selection bias exists. Another potential threat is the history effects. For instance, when it is difficult to separate out how much of the increase in the output was due to the experimental conditions, and how much was due to sample participants' attributes, the internal validity of the research is low.

4.4.2.3   Construct validity

When examining the construct validity, we test whether the results represent what is theoretically intended. It addresses the following question 'Am I really measuring the right construct that was intended theoretically?' For example, if a researcher wants to know a particular design strategy (variable A) will be effective for improving product quality (variable B), he or she will need at least one measure of product quality. If that measure does not truly reflect product quality levels but rather, for instance, product development time (confounding variable X), then the study will be lacking construct validity, and the operationalisation of the product quality construct is low, that is, it does not represent what is theoretically intended.

4.4.2.4   External validity
External validity addresses the issue of being able to generalise the results of a study to a wider scope. An example of the threats to the external validity is the following question: 'Would I find these same results with a different sample?' If the researcher cannot answer 'yes' to this question, then the external validity of the study is threatened.

4.5   Recommended research methodology

The main goal of this book is to help in identifying the engineering design activities and strategies that are critical to the commercial success of the product and to differentiate successful from unsuccessful products, that is, the product success differentiators among design activities and strategies. This book is affiliated with the intersection between Modelling the Design Process, which provides answers to 'What-if' questions, and Integrated System Design, which identifies the best route through the design process. The proposed research methodology in this book is based on a research experimental approach through conducting two design research experiments. The design research experiments are based on a first-year engineering undergraduate structural design project and a second-year engineering undergraduate integrated design project. Both of these two projects are run at the Department of Engineering, Cambridge University; therefore they are easily accessible to researchers. They collectively reflect much of the real-world engineering design environment.

This book adopts hybridisation of both product-oriented design research and process-oriented design research perspectives to cost-effectively improve product performance attributes, and consequently to make the product more successful in the market and eventually to improve the industrial companies' profitability. In addition, the hybrid approach to understanding and managing the product development process is employed in this book, as it represents the most comprehensive approach in this context. In addition, the research methodology used in this work adopts hybridisation of all the three major research methods: exploratory (causal), descriptive research and correlational research as the research needed some tools of each of these. For instance, from exploratory research, the quasi-experiment was used; from descriptive research, frequency analysis was needed and from correlational research, the causal comparative observational research and archival research were employed. Moreover, this book adopts the hybrid quantitative–qualitative research approach because the research needed some tools of both; for instance, from quantitative research, questionnaires and statistics were employed, and from qualitative research, observational research was used. This book implements the statistical analysis approach to rigorously analyse the research data and verify and validate the research results.

The research phases adopted in this book are (1) setting research goals and criteria, (2) conducting literature review and accordingly identifying research questions and hypotheses, (3) collecting data empirically and conducting statistical analysis and (4) verifying and validating the research results. In the studies described in this book, archival research that is part of a correlational study was conducted first. Next, observational research was conducted as part of a correlational study. Next, quasi-experiments were conducted, which are part of an exploratory study. Then, survey research was conducted, which is part of a correlational study, followed by a descriptive study. Next, causal comparative research was conducted, which is part of correlational research. In this book two types of correlational studies are described: bivariate correlation and the cross-lagged panel design for inferential causality. Finally, industrial case studies, which are part of an exploratory study, are planned to be conducted for further verification.

The proposed research methodology in this book consists of the following steps:

1. Conducting a research literature review, which constitutes the secondary source of data in this work, to identify the research gap and primary research questions, and consequently primary research hypotheses; in addition, the researcher obtains secondary research hypotheses from the literature.
2. Conducting two research experiments from which the researcher gains experimental observations that constitute the first part of the primary source of data in this book.
3. Based on the secondary research hypotheses in the literature, the researcher constructs experimental questionnaires.
4. Based on both the experimental observations and responses to the experimental questionnaires that constitute the second part of the primary source of data in this book the researcher statistically tests, using statistical package for social sciences (SPSS), the correlation between the literature secondary research hypotheses and the experimental observations on the one hand and the product success on the other.

5. Testing causality of the experimental research results using the inferential statistics.
6. Based on the statistical analysis of the experimental data, the researcher identifies a set of experimentally valid important design activities and strategies.

To implement that research methodology, data manipulation and formulation and statistical analysis were conducted.

4.5.1 Data formulation and manipulation

The data formulation included implementation of the rules mentioned in the following sections: sampling, that is, Section 3.5, questionnaires data collection method, that is, Section 3.7.2 and observational surveys data collection method, that is, Section 3.7.3. The data manipulation included the following tasks:

- Handling blank responses to the questionnaire
- Handling the Do Not Know responses from the respondents to the questionnaire
- Handling inversed questions in the questionnaire

4.5.2 Statistical analysis approach

This book proposes a statistical approach that uses both descriptive statistics and inferential-statistics. The collected data is organised in a categorical interval-scale fashion. The nonparametric statistic will be adopted using SPSS in the inferential-statistics part to avoid making assumptions about the population's parameters and consequently to improve the validity of the statistical analysis results. The proposed statistical approach has the following five attributes:

1. Population: In the two experiments that were conducted, there was one population that is the complex products' engineering designers population; the term 'complex products engineering designers' in this work refers to the people who design complex engineering products for commercial purposes or for academic research purposes; also the term 'complex product' refers to the product that either exhibits two or more engineering disciplines in the design process or necessitates engagement of two or more engineering designers in the design process.
2. Sampling frame: A bounded and unbiased collection of the targeted individuals in which every single individual is identified and can be selected.

3. Data type: Categorical random variable (rather than continuous numerical).
4. Sampling design: The adopted sampling design was probability simple random sampling because of its cost-effectiveness and reasonable accuracy.
5. Sample size and number of samples: The target sample size was 30, as the minimum statistically representative sample size (Alder and Roessler, 1962; Wackerly et al., 1996).

The adopted statistical approach consists of the following steps (Alder and Roessler, 1962; Wackerly et al., 1996):

1. Identifying the dependent variable(s) and the independent parameter(s); for instance, the dependent variable is the product success, and the independent parameters are the experimentally hypothesised design activities and strategies.
2. Constructing the theoretical framework, that is, rational expectations and relationships between the dependent variable and the independent parameters.
3. Identifying the primary and secondary hypotheses.
4. Constructing a questionnaire that stresses the research hypotheses.
5. Running descriptive statistics, such as central tendencies, dispersion measures.
6. Analysis of data using the Spearman rank correlation coefficient, that is, implications of the measurement and testing of the hypotheses.
7. Generalisation, which includes significance level, type I and type II errors, and variation in the dependent variables based on the correlation coefficient.

chapter five

Design experiments
Description and outcome

The term 'experiment' usually implies a controlled experiment, but some-times controlled experiments are prohibitively difficult or impossible, such as in the real engineering design process. In this case researchers resort to quasi-experiments, which implies that these are uncontrolled experiments (Sekaran, 2003; Robson, 2002). Quasi-experiments rely mainly on observation of the parameters of the system under study, rather than on manipulation of a few variables as occurs in controlled experiments. To the degree possible, they attempt to collect data for the system in such a way that contribution from all variables can be determined, and where the effects of variation in certain variables remain approximately constant so that the effects of other variables can be discerned.

The statistical analysis in this book is based on six samples that were drawn from a target population, three of which came from the Integrated Design Project (IDP) and the other three from the Structural Design Project (SDP). The total number of designers in the IDP experiment was 174 with a sample size of 58. The total number of designers in the SDP experiment was 91 with a sample size of 30. The target population of this analysis is the engineering designers in both the robotics and flexible manufacturing systems and the construction and structural design industrial sectors. Responses of the designers to the questionnaires that included both the experimental secondary hypotheses and experimental observations were collected. The collected data were analysed using both descriptive and inferential statistics. In the descriptive statistics, a frequency analysis of the data was conducted including the mean and standard deviation. In the inferential analysis, a nonparametric statistical analysis was conducted using the Spearman correlation coefficient to avoid making any assumption about the population parameters. This approach provides more rigourous results than parametric statistical analysis (Wackerly et al., 1996). This chapter presents the outcome of the design quasi-experiments.

5.1 Quasi-experiment research

This book presents two research quasi-experiments in which engineering design students were observed while designing products and were respondents to questionnaires, which are presented in Appendices A and B; the first one of these two experiments and questionnaires is based on the IDP whereas the second is based on the SDP (Cambridge University, Department of Engineering [CUED], 2007a,b). The design independent variables in these experiments were the design activities and strategies, and the design dependent variables were the product's features, performance and market success (Elmoselhy, 2014).

In this context and for clarification purposes, the definition of the term 'Observation' is an occurrence of a design activity or implementation of a design strategy, which was observed or identified while observing the design process. Also, the term 'Hypothesis' is defined as a postulated design activity or strategy that was identified by reviewing the relevant engineering design literature, but the observer was not sure to what degree that design activity occurred or what strategy was implemented, as there were times during which the observer had no access to the design process. The SDP and the IDP are chosen to conduct a quasi-experiment on each of them. In both of these two experiments, data were collected from observations and questionnaires. The product performance evaluation criteria of the SDP are product strength, weight and cost. In the IDP, the product performance evaluation criteria are collectively the robot performance score on the competition day based on performing specific tasks, within specific space limitations and within a specific timeframe. This section presents details of these two quasi-experiments.

5.1.1   Structural Design Project

The SDP has specific design features and constraints and specific success criteria. These design features and constraints and success criteria are elaborated in Sections 5.1.1.1 and 5.1.1.2.

5.1.1.1   SDP features and constraints

The design features and constraints of the SDP are (CUED, 2007a)

1. Designers work to design, build and test to destruction a lightweight truss with a span of about 1 m.
2. Each design team produces a set of working drawings for its structure with a structure's cost sheet. These drawings are handed in during fabrication time in the workshop.
3. Each design team has a month to design its structure and 15 hours, in terms of five sessions, to manufacture and build its structure.
4. Each structure is finally weighed and tested to destruction.

5.1.1.2   SDP success criteria

In the SDP, a cantilever-like structure is required to carry a vertical load W at a horizontal distance of 815 mm from a rigid vertical plate as shown

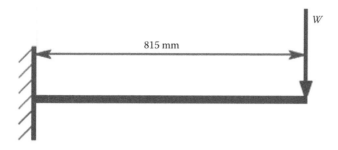

Figure 5.1 Structural Design Project problem. (From Cambridge University, Engineering Department (CUED). (2007a). *First Year Undergraduate Structural Design Project.* Cambridge, UK: Cambridge University Press.)

in Figure 5.1. The plate has four pairs of M6 tapped holes to which the structure may be attached. The load is applied to the structure through a spreader bar of length 115 mm, thus dividing the applied load into two loads, W/2.

The working loads are $W = +1350$ N and $W = -135$ N; at both of these loads, there must be no visible deformation. The load factor at collapse equals 2, according to the applicable standard collapse load, and is applicable to both positive and negative values of W. It is essential to inspect the loading arrangements before detailed drawings are made; the designers are responsible for making any required measurements. The objective is to design a structure with the given material that will satisfy the loading conditions; importantly, the structure must be lightweight and simply made. The following formula was used to work out the performance index of design based on structure weight, structure cost and structure failure load.

$$\text{Normalised Score} = \left[-(A) \times \frac{\text{Structure's Weight} - \text{Average Weight}}{\text{Average Weight}} \right]$$

$$- \left[(A) \times \frac{\text{Structure's Total Cost} - \text{Average Cost}}{\text{Average Cost}} \right]$$

$$+ \left[\frac{\text{Observed Collapse Load}}{\text{Standard Collapse Load}} \right] \tag{5.1}$$

where

$$A = 1, \quad \text{if} \quad \frac{\text{Observed Collapse Load}}{\text{Standard Collapse Load}} > 1$$

$$A = \frac{\text{Observed Collapse Load}}{\text{Standard Collapse Load}}, \quad \text{otherwise}$$

5.1.2   Integrated Design Project

The IDP has specific design features and constraints and specific success criteria, which are elaborated in Sections 5.1.2.1 and 5.1.2.2.

5.1.2.1   IDP features and constraints

Modern design often requires an integrated system to be designed by a multidisciplinary team, as reflected by this design, build and test project. The disciplines represented in this project are mechanical, electronics and software engineering. This project involves designing and building a model autonomous guided vehicle (AGV) or mobile robot from a kit of parts. It is a mechanical device with associated electronic circuits and control software. The combination is often categorised by the term 'mechatronics' (CUED, 2007b). Credit is given for the project management skills in terms of planning and execution of the design and construction process and system design skills in terms of quality of the complete vehicle. A competition is held shortly after the four-week project period to find the best robot in terms of meeting the design specification and speed of performance.

5.1.2.2   IDP success criteria

Each design team is divided into three subteams, with one subteam responsible for the mechanical aspects, one for the electronics and one for the software. The entire team is responsible for the integration of these parts to form the complete AGV. The task is to design and build a mobile robot as an AGV that collects six 'eggs' from a conveyer belt (B) and transports them to one of two delivery points depending on the type of egg within five minutes. Rubber eggs should be delivered to the bin at D1 and chocolate eggs placed in the egg box on the lorry at D2, as shown in Figure 5.2. The task will continue until half a dozen eggs have been transferred or the time limit of five minutes is reached. All movements are to be carried out within the playing area of 2400 mm × 2400 mm. The playing area's dimensions and conditions are (1) pallet dimensions of 100 × 100 mm, (2) pallet weight of 80 g, (3) rubber egg's weight of 65 g, (4) chocolate egg's weight of 40 g and (5) conveyer speed of 10 cm/s. In addition, the following conditions are applied:

- The 'eggs' will be carried on small pallets while on the conveyer belt.
- The conveyer can be started/stopped and reversed using optical switches, light-dependent resistor (LDR) based.
- An adjustable light-emitting diode (LED) beam, suitable for driving an optical sensor, is available mounted just above the conveyer belt.

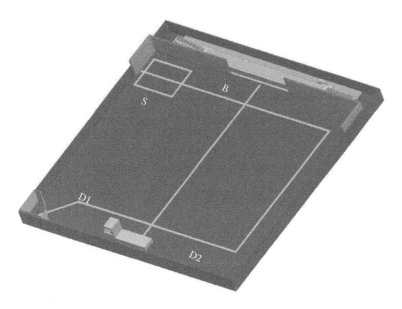

Figure 5.2 Integrated Design Project playing area conditions. (From Elmoselhy, S. A. M. (2015). "Empirical investigation of a hybrid lean-agile design paradigm for mobile robots." *Journal of Intelligent Systems*, 24(1).)

5.1.3   Constraints on the conducted experiments

The following system design constraints were imposed on the two conducted experiments:

1. There were some system design constraints imposed on the IDP such as not to handle controllability, for example, overshooting and undershooting, using electronic components such as logic gates; although handling controllability by electronic components, such as programmable logic controller, is more responsive than by software.
2. There were some system design constraints imposed on the SDP such as using only aluminium alloy as the structural material to facilitate prototype manufacturing, building and processing.

5.2   Quasi-experiments research outcome

The conducted quasi-experiments resulted in five main research outcomes: (1) experimentally observed common design activities and strategies in the SDP, (2) experimentally observed common design activities and strategies in the IDP, (3) observed reasons for rework in the two implemented design projects, (4) experimental research observations and (5) responses to the questionnaires.

5.2.1 *Experimentally observed common design activities and strategies in the SDP*

In the SDP, it was observed that there are design activities and strategies that are commonly conducted and adopted by the designers. Throughout the entire SDP, from the scoping phase, to the conceptual design phase, to the preliminary design phase, to the detailed design phase, to the to-scale prototype construction phase, the following observed design activities and strategies were common among designers.

5.2.1.1 *Scoping phase*

The observed common design activities and strategies in the scoping phase of the SDP were

1. Team organisation
2. Task clarification
3. Delivering requirements specification

5.2.1.2 *Conceptual design phase*

The observed common design activities and strategies in the conceptual design phase of the structural design project were

1. Analysing the requirements specification; for instance, how is the structure likely to fail? A likely answer to this question is compression members buckling.
2. Setting a design strategy; for instance, minimising length of compression members
3. Generating design ideas; for instance, sketching design ideas that include the set design strategy

5.2.1.3 *Preliminary design phase*

The observed common design activities and strategies in the preliminary design phase of the SDP were

1. Load analysis
2. Structure member design; for instance, how large does each member need to be when using a triangulated structure?
3. Joint design; for instance, how many rivets are required on each member?

4. Adopting rivets to fasten members if the carried load is pure shear and adopting bolts otherwise, for example, tension, compression or combined load
5. Bracing member design

5.2.1.4 Detailed design phase
The observed common design activities and strategies in the detailed design phase of the structural design project were

1. Designing each vertical plane first
2. Working out details of joints
3. Making sure that the relevant lines of action meet up at a single point
4. Working out details of bracing members
5. Working out details of the structure layout
6. Working out details of the structure members

5.2.1.5 To-scale prototype construction phase
The observed common design strategy in the to-scale prototype construction phase of the structural design project was as follows:

1. Making to-scale prototype of the detailed design structure of soft material, for example, aluminium, to facilitate the manufacturing process to make the required structure if a mistake occurred.

5.2.2 Experimentally observed common design activities and strategies and typical design task allocation in the IDP

In the IDP, it was observed that there are design activities and strategies that are commonly conducted and adopted by the designers. These common design activities and strategies observed throughout the entire IDP, from the scoping phase, to the conceptual design phase, to the preliminary design phase, to the detailed design phase, to the to-scale prototype construction phase, are presented Sections 5.2.2.1 and 5.2.2.2. In addition, the typical design task allocation for each subgroup in the IDP robot design project is presented. Finally, the inputs to and outputs from the robot, as a system, are presented.

5.2.2.1 Experimentally observed common design activities and strategies in the IDP
In the IDP, it was observed that there are design activities and strategies that are commonly conducted and adopted by the designers. Throughout the entire IDP, from the scoping phase, to the conceptual design phase, to the preliminary design phase, to the detailed design phase, to the to-scale

prototype construction phase, the following observed design activities and strategies were common among designers.

5.2.2.1.1 Scoping phase The observed common design activities and strategies in the scoping phase of the IDP were

1. Team organisation
2. Task clarification
3. Delivering requirements specification

5.2.2.1.2 Conceptual design phase The observed common design activities and strategies in the conceptual design phase of the IDP were

1. Brainstorming solutions
2. Delivering preliminary design concepts

5.2.2.1.3 Preliminary design phase The observed common design activities and strategies in the preliminary design phase of the IDP were

1. Setting interface specifications
2. Evaluating design concepts
3. Delivering the final concept
4. Making a progress presentation in a design review meeting

5.2.2.1.4 Detailed design phase The observed common design activities and strategies in the detailed design phase of the IDP were

1. Mechanical detailed design
    1.1 Working out mechanical detailed design
        i. Mechanical design layout
        ii. Chassis
        iii. Lift arm
        iv. Drive system
        v. Sensor mountings
    1.2 Delivering mechanical drawings for final acceptance of design
2. Electronics detailed design
    2.1 Working out layout of electronic components
    2.2 Working out detailed electronic circuits of the electronic components, for example, opto-sensors, strain gauges and microswitches
    2.3 Actuator drive
    2.4 Interfacing printed circuit boards (PCBs)
    2.5 Delivering schematic diagrams of electronics for final acceptance of design

3. Software detailed design
    3.1 Conducting components testing
    3.2 Working out logic structure to realise the interface specifications, for example, working out route planning and deciding which turn to make, deciding when to make turn and move arm, deciding when to move wheels to follow lines
    3.3 Generating software code for final acceptance of design

5.2.2.1.5   Full-scale prototype construction phase   The observed common design activities and strategies in the full-scale prototype construction phase of the integrated design project were

1. Mechanical construction
    1.1 Mechanical design layout
    1.2 Chassis
    1.3 Lift arm
    1.4 Drive system
    1.5 Sensor mountings
    1.6 Mechanical assembly
2. Electronic circuit construction
    2.1 Constructing electronic circuits of the electronic components, for example, opto-sensors, strain-gauges and micro-switches
    2.2 Constructing electronic circuits of the actuator drive
    2.3 Constructing interfacing PCBs
    2.4 Assembling electronic circuits together
    2.5 Conducting circuit testing, for example, observing proper working of LEDs as indicators while applying voltage
3. Software implementation
    3.1 Implementing code for the planned route
    3.2 Implementing code for line following in the field
    3.3 Implementing code for controlling and synchronising the drive system
    3.4 Implementing code for load sensing and identification
    3.5 Implementing code for load delivery
    3.6 Testing code and making modifications accordingly
4. Making a final presentation in a design review meeting

5.2.2.2   Typical design task allocation for each subgroup in the IDP
The typical design task allocation for each subgroup in the IDP, robot design project, is presented in this section. This covers the three main subsystems of robot design: mechanical, electronics and software subsystems.

5.2.2.2.1   Mechanical subsystem design subgroup   The typical design tasks allocated for the mechanical subsystem design subgroup are

1. Chassis design and construction
2. Lift mechanism design and construction

5.2.2.2.2   Electronics subsystem   The typical design tasks allocated for the electronics subsystem design subgroup are

1. Sensor type and specification
2. Load detection and accordingly classification
3. Sensor/software interface
4. Pneumatic actuator control circuits

5.2.2.2.3   Software subsystem   The typical design tasks allocated for the software subsystem design subgroup are

1. Setting a strategy for both route planning and sensor positioning
2. Control of movement, direction and line following based on sensor information
3. Motor testing and calibration

5.2.2.3   Inputs to and outputs from robot as a system
The observed inputs to and outputs from robot, as a system, are key to understanding the integrated design system. These observed inputs to and outputs from robot are presented in this subsection.

5.2.2.3.1   Inputs to robots   The observed inputs to robot as a system are

1. Digital signals from microswitches, to avoid collision with obstacles and to detect reaching the load base
2. Digital signals from light sensors, such as LDRs, for line following and for colour discrimination
3. Analogue signals from strain gauges, to measure weight of the load

5.2.2.3.2   Outputs from robots   The observed outputs from robots as a system are

1. Signal to direct current (DC) motor, with LED as an error-tracing element
2. Signal to solenoid valves of pneumatic actuators

5.2.3   *Reasons for rework in the two design projects*

In both of these design projects, it was observed that there are specific reasons for rework:

1. To save weight, assembly time, and/or cost
2. To avoid unexpected mechanisms of failure or of motion
3. To correct an incorrect assembly, or to correct an unexpected path in the robot project

5.2.4   *Experimental research observations*

The research observations obtained from examining the design activities of the engineering design students in the SDP and IDP are presented in Table 5.1. Wherever the statement of 'In the integrated design system' is found in this book, it implies that point is relevant only to the IDP rather than to the SDP.

5.2.5   *Responses to the questionnaires*

The responses to the questionnaire were categorised as follows:

Strongly disagree = 1
Disagree = 2
Agree = 3
Strongly agree = 4

The design activities and strategies in the design experiments are presented in Appendix C. Any value in the data set other than 1, 2, 3, and 4 is the average of respondents' responses in that column. The average value in each column is used to fill in the gap of empty responses as a way of manipulating responses to questionnaires (Sekaran, 2003). To realise the target sample size of 30, partial samples were sought. Six partial samples could be realised, three of which were drawn from the IDP experiment with a partial sample size of 58 each, and the other three samples from the SDP experiment with a partial sample size of 30 each. The total sample size that could be realised was therefore 265, which meets the target sample size. This satisfies the first of the two criteria of the sample representativeness: sample size and sampling design. The second criterion of sample representativeness, sampling design, was also satisfied because the sampling design used in this book was based on probability simple random sampling, which is suitable for generalisation purposes and cost-effectively results in reasonably fair results. Although the sample size of

Table 5.1 Research observations

Research observations	Sub-system type	Design phase	Design strategy/ activity	Operationalisation and mapping of questionnaire's questions
System design (SD) dimension				
1. In the integrated design system, those who shifted complexity towards the software subsystem, rather than the mechanical subsystem, had designs that were less vulnerable to failure modes (e.g. out-of-plane buckling, roll mode and pitch mode instability and friction wear) (and consequently had better reliability) and less exposed and less sensitive to the uncontrollable external factors (and consequently had better robustness) and had the successful designs.	SD	Conceptual and preliminary design phases	Strategy	Question 1
2. In the integrated design system, iterations in the software subsystem exhibited shorter development time (and consequently incurred lower development cost) in comparison with the mechanical subsystem counterparts.	SD	Preliminary and detailed design phases	Activity	Question 18

(Continued)

Table 5.1 (Continued) Research observations

Research observations	Sub-system type	Design phase	Design strategy/ activity	Operationalisation and mapping of questionnaire's questions
3. In the integrated design system, the constraints in the virtual space, for example, data processing time, are fewer in comparison with the constraints in the physical space; for example, suspension arrangements and space required for a mechanical distinguishing, that is, sensing, device are more in comparison with those required for a software-controlled electronic device. Consequently, shifting design complexity towards the software subsystem provided more flexibility and adaptability to the design.	SD	Conceptual and preliminary design phases	Strategy	Question 1
4. Those who adopted the simplest design, which addresses the design problem requirements specification by the minimum set and most effective combination of system components, rather than the cheapest and the lightest weight design, had the most successful designs on the total success criteria.	SD	Conceptual and preliminary design phases	Strategy	Question 2
5. The most constrained design subsystem across the integrated design system is the electronics subsystem.	SD	Conceptual design phase	Activity	Question 3

(Continued)

Table 5.1 (Continued) Research observations

Research observations	Sub-system type	Design phase	Design strategy/activity	Operationalisation and mapping of questionnaire's questions
6. In the integrated design system, the mechanical design concept had been chosen first, then the electronic design concept, as the mechanical concept dictates the electronic interface; then the software design concept was set to enable the mechanical concept to perform the required function.	SD	Conceptual design phase	Activity	Question 3
7. In the integrated design system, adopting electronic distinguishing, that is, sensing, components, whenever they are applicable and viable options, to distinguish the loads that were required to be transported, rather than comparable mechanical distinguishing systems because of better reliability, in terms of accuracy and lower vulnerability to failure modes (e.g. out-of-plane buckling, roll mode and pitch mode instability and friction wear), better robustness, in terms of being less exposed and less sensitive to the uncontrollable external factors and better flexibility, in terms of wide bandwidth, of the former, improved reliability and agility of the overall system.	SD	Conceptual, preliminary and detailed design phases	Strategy	Question 4
8. In the integrated design system, those designers who adopted nonstandard components ended up with unreliable and failed designs.	SD	Preliminary and detailed design phases	Strategy	Question 5

(Continued)

Table 5.1 (Continued) Research observations

Research observations	Sub-system type	Design phase	Design strategy/ activity	Operationalisation and mapping of questionnaire's questions
9. In the integrated design system, adopting a tolerant design that can accommodate variations in the inputs to the system, to improve robustness of the robot final concept, improves the opportunity of achieving a better design.	SD	Conceptual and preliminary design phases	Strategy	Question 10
10. In the integrated design system, shifting complexity towards the software, rather than the mechanical subsystem, leads to the largest number of design iterations, if any, within the software subsystem, followed accordingly by the number of iterations in the electronics subsystem.	SD	Preliminary and detailed design phases	Activity	Question 18
Project management (PM) dimension				
11. In the integrated design system, those designers who did not conduct early verification and validation of the design concept, for example, early testing in the design process, became trapped in incompetent design concepts, experienced a large unfavourable impact of rework and consequently had longer development time, which implies larger development cost, and received a low product performance score.	PM	Conceptual and preliminary design phases	Activity	Question 19

(Continued)

Table 5.1 (Continued) Research observations

Research observations	Sub-system type	Design phase	Design strategy/activity	Operationalisation and mapping of questionnaire's questions
12. In the integrated design system, selecting a team leader with a software specialisation and familiarity with electronics and mechanical subsystems led to a higher performance score than otherwise. There could be two reasons for this observation. First, the software subsystem is intangible and difficult to be conceived by unspecialised designers. Second, software subsystem design effectively starts after the mechanical and electronics design concepts are decided on and consequently the software designer has a better opportunity to see and manage the overall project.	PM	Scoping phase	Activity	Question 20
13. At the same level of designers' effort and for the same type of design projects, spending more time on the conceptual and preliminary design phases together than on the detailed design phase improves the opportunity of coming across a better design.	PM	Conceptual, preliminary and detailed design phases	Activity	Question 21

(Continued)

Table 5.1 (Continued) Research observations

Research observations	Sub-system type	Design phase	Design strategy/ activity	Operationalisation and mapping of questionnaire's questions
14. In the integrated design system, assigning at least 5% of the total time duration of the project to a macroscale aggregate plan in the scoping phase and having resource provisions, for example, time buffer, for variations in details and making a detailed design plan at the preliminary design phase improve the agility of the design process.	PM	Scoping and preliminary design phases	Activity	Question 24
15. Making more iterations in the conceptual and preliminary design phases than in the detailed design phase improves the opportunity for achieving a better design, while minimising waste of time and resources.	PM	Conceptual, preliminary and detailed design phases	Activity	Question 28
16. In the integrated design system, the design team estimates the duration of each project activity based on work breakdown structure, experience and comparison with comparable projects.	PM	Scoping phase	Activity	Question 34
17. In the integrated design system, the design subteams coordinate with each other concerning the interrelated elements of the system, for example, software–electronics interconnection for microchip pins allocation and for positioning of sensors that fits all the required routing.	PM	Preliminary and detailed design phases	Activity	Question 37

(Continued)

Table 5.1 (Continued) Research observations

Research observations	Sub-system type	Design phase	Design strategy/ activity	Operationalisation and mapping of questionnaire's questions
18. In the integrated design system, the total time duration of the project is the most important constraint, rather than a resource, as designers cannot control the rate at which it is expended and imposed on the project, and those who did not manage their time properly, that is, did not start their design process as early as possible in the project time frame and interrupted their design process by not assigning sufficient time for presenting the outcome of the design process milestones did not allocate time appropriately for each design phase and consequently ended up with an ineffective design concept.	PM	Scoping and conceptual design phases	Activity	Question 39
19. In the integrated design system, the design team that had all of its design subteams meet up collectively in the conceptual design phase to choose the mechanical and electronics design concepts, respectively, attained a higher product performance score.	PM	Conceptual design phase	Activity	Question 45

(Continued)

Table 5.1 (Continued) Research observations

Research observations	Sub-system type	Design phase	Design strategy/ activity	Operationalisation and mapping of questionnaire's questions
20. In the integrated design system, the design team that had its electronics and software design subteams conduct initial test experiments in the preliminary design phase on the chosen conceptual, mechanical and electronics design components to explore the full potential and scope of these components achieved a more effective design than otherwise and a higher product performance score.	PM	Preliminary design phase	Activity	Question 46
Mechanical (M) subsystem design dimension				
21. Those who constructed a to-scale cardboard prototype, mock-up or equivalent in the mechanical subsystem attained a higher performance score.	M	Preliminary design phase	Activity	Question 47
22. Adopting a strength-adaptable mechanical chassis design approach that can embrace further changes, for example, future changes in weight and layout of electronic components, to improve agility of the design process, improved the designers' opportunities for achieving a better design.	M	Conceptual, preliminary and detailed design phases	Activity	Question 48

(Continued)

Table 5.1 (Continued) Research observations

Research observations	Sub-system type	Design phase	Design strategy/ activity	Operationalisation and mapping of questionnaire's questions
23. Making a prediction of the progressive failure of the design through drawing sketches or through conducting a finite element analysis, to avoid unexpected failure, improved the opportunity for achieving a better design.	M	Conceptual and preliminary design phases	Activity	Question 49
24. Making sketches to generate concepts in the conceptual design phase improved the opportunity for achieving a better design.	M	Conceptual design phase	Activity	Question 50
25. Constructing computer-aided design (CAD) models in the preliminary design phase improved the opportunity for achieving a better design.	M	Preliminary design phase	Activity	Question 51
26. Building a to-scale design prototype to be tested in the real operational environment in the detailed design phase improved the opportunity for achieving a better design.	M	Detailed design phase	Activity	Question 52
27. In the integrated design system, the mechanical subteam determines the optimal number of degrees of freedom of the system, for example, of picking up and unloading mechanisms, that fulfil the functional requirements and minimise the number of needed controllable mechanisms.	M	Preliminary and detailed design phases	Activity	Question 55

(Continued)

Table 5.1 (Continued) Research observations

Research observations	Sub-system type	Design phase	Design strategy/ activity	Operationalisation and mapping of questionnaire's questions
Electronics (E) subsystem design dimension				
28. In the integrated design system, those who set an error-tracing operational design strategy, such as coloured electronic wiring in the electronics subsystem, to facilitate iterations, troubleshooting, software code debugging and/or maintainability of the product final concept, attained a higher performance score.	E	Preliminary and detailed design phases	Strategy	Question 56
29. In the integrated design system, the electronics designers choose the most reliable types of sensors and extend the sensors' robustness through software manipulation.	E	Conceptual, preliminary and detailed design phases	Activity	Question 57
30. In the integrated design system, the electronics designers develop the electronics circuitry such that it minimises the number of required reads and writes and maximises utilisation of channelling and port addressing.	E	Conceptual, preliminary and detailed design phases	Activity	Question 58

(Continued)

Table 5.1 (Continued) Research observations

Research observations	Sub-system type	Design phase	Design strategy/ activity	Operationalisation and mapping of questionnaire's questions
Software (S) subsystem design dimension				
31. In the integrated design system, spending at least 10% of the total time duration of the software subsystem on functional analysis ard software architecture, for example, making a program flow chart, rather than on the programming code implementation itself, to minimise waste of time and resources and to eliminate sources of avoidable mistakes, improved designers' opportunities for achieving a better design.	S	Scoping phase	Activity	Question 59
32. In the integrated design system, assigning at least 10% of the total time duration of the software subsystem to the overall testing time, that is, to both quick testing and extensive testing, improved designers' opportunities for achieving a better design.	S	Scoping phase	Activity	Question 61

(Continued)

Table 5.1 (Continued) Research observations

Research observations	Sub-system type	Design phase	Design strategy/ activity	Operationalisation and mapping of questionnaire's questions
33. In the integrated design system, testing aggregately the software subsystem at the end of the project rather than adopting quick testing lessens designers' opportunities for achieving a better design.	S	Scoping phase	Activity	Question 62
34. In the integrated design system, adopting effective programming techniques, such as selective switch case rather than enumerated switch case in the software programming code implementation, to minimise waste in the design development time, improved the opportunity for achieving a better design.	S	Preliminary and detailed design phases	Activity	Question 63

30 can be accepted as the minimum statistically representative sample size according to Alder and Roessler (1962) and Wackerly et al. (1996), Krejcie and Morgan (1970) argued that the sample size for a large population size of 100,000+, which might be the case in the complex products' engineering designers' populations, should be 384. The difference between such a sample size and the total sample size that could be realised could constitute a probability of error in generalising the sample results to the population because such a difference might affect the sample representativeness. Because the sample representativeness depends on two factors – sampling design and sample size – the probability of error in sample results generalisation consequently depends on these two factors. Therefore, such a probability of error in generalising the sample results can be estimated using the following formula:

$$
\begin{aligned}
&\text{Probability of no error in generalising sample results} \\
&= \text{Probability of meeting the sample size criterion} \\
&+ \text{Probability of meeting the sampling design criterion} \quad (5.2)
\end{aligned}
$$

Thus,

$$
\text{Probability of no error in generalising sample results} = \left(0.5 - \frac{384 - 265}{384} \right)
$$

$$
+\, 0.5 = 0.85
$$

$$
\begin{aligned}
\text{Probability of error in generalising sample results} = 1 - &\text{Probability of no} \\
&\text{error in generalising} \\
&\text{sample results} \quad (5.3)
\end{aligned}
$$

Hence,

$$
\text{Probability of error in generalising sample results} = 0.15
$$

Given this result of the probability of error in generalising sample results of this work, the probability of no error in generalising sample results is almost sixfold the probability of error in generalising sample results, which proves the rigour of the statistical results.

chapter six

Statistical analysis results on the experiments' outcomes

The statistical analysis in this book is based on six samples that were drawn from a target population, three of which came from the Integrated Design Project (IDP) and the other three samples from the Structural Design Project (SDP). The total number of designers in the IDP experiment was 174 with a sample size of 58. The total number of designers in the SDP experiment was 91 with a sample size of 30. The target population of this analysis is the engineering designers in both the robotics and flexible manufacturing systems and the construction and structural design industrial sectors. Responses of the designers to the questionnaires that included both the experimental secondary hypotheses and experimental observations were collected. The experimental research questionnaires are presented in Appendices A and B. Accordingly, the collected data were analysed using both descriptive and inferential statistics. In the descriptive statistics a frequency analysis of the data was conducted including the mean and standard deviation. In the inferential analysis, a nonparametric statistical analysis was conducted using a Spearman correlation coefficient to avoid making any assumption about the population parameters. This approach provides more rigourous results than the parametric statistical analysis (Wackerly et al., 1996). This chapter presents the statistical analysis results using SPSS. It presents as well the verification and validation of the quasi-experiments. The implication of percentage of variation in technical performance due to the design variable (r^2) is also presented.

6.1 Results of analysis of bivariate correlation with the technical attributes of product success

Tables 6.1 to 6.3 present the statistical analysis results of the bivariate correlation analysis in descending order. The ranges of statistical correlation adopted in this book are

1. No correlation when the correlation coefficient ranges from 0 to less than 0.1
2. Low correlation when the correlation coefficient ranges from 0.1 to less than 0.3

Table 6.1 Positively correlated design activities and strategies with the technical attributes of product success

Design activities/strategies		Percentage of variation in mobile robot performance (r^2)
18-ACT	Largest number of iterations in software	0.3**
1-STRA	Shifting complexity	0.2**
29-ACT	Modular testable interdeliverables	0.2**
14-STRA	Considering reliability	0.2**
19-ACT	Early verification	0.1**
13-STRA	Fast response versus stability, accuracy and payload	0.1**
36-ACT	Modular deliverables and testing	0.1**
26-STRA	Resource provisions	0.1**
25-ACT	Multidisciplinary team	0.1**
27-STRA	Modular testing	0.1**
31-STRA	Decentralised decision making	0.1**
39-ACT	Starting the design process as early as possible and assigning time for presenting the outcome	0.1**
41-ACT	Documenting the outcome of design discussions	0.1**
44-ACT	Meeting collectively in conceptual rather than in detailed design	0.1**
49-ACT	Predicting design progressive failure	0.1**
54-ACT	Checking accuracy of manufacturing	0.1**
60-ACT	Quick testing of interdeliverables between modules	0.1**
2-STRA	Simplest design	0.1**
12-ACT	Top-down system structure decomposition	0.1**

** Correlation is significant at the 0.01 level (2-tailed).

Table 6.2 Negatively correlated design activities and strategies with product success technical attributes

Design activities/strategies	Percentage of variation in mobile robot performance (r^2)
62-ACT Testing aggregately the software subsystem at the end of the project rather than adopting quick testing	−0.3**

** Correlation is significant at the 0.01 level (2-tailed).

Table 6.3 Uncorrelated design activities and strategies with the technical attributes of product success

Design activities/strategies	
3-ACT	Choosing the mechanical design concept first
4-STRA	Electronic distinguishing
5-STRA	Only standard components
6-STRA	Modular design
7-ACT	Systematic design approach
8-ACT	Quality of performing systematic design approach
9-STRA	Error proof
10-STRA	Tolerant design
11-STRA	Maximising functionality and minimising iterations
15-STRA	Considering maintainability
16-STRA	Considering safety
17-STRA	Error recovery
20-ACT	Software team leader
21-ACT	More time on conceptual and preliminary phases
22-ACT	Reviewing similar designs
23-ACT	Cross fertilisation of concepts among the team
24-ACT	Time to aggregate plan
28-ACT	More iterations in conceptual and preliminary phases
30-ACT	Live meetings team in the conceptual and preliminary phases
32-STRA	Sequential and concurrent design tasks
33-STRA	More interested in meeting delivery deadline than in cost
34-ACT	Estimating design activity duration
35-STRA	Redeployment of human resources
37-ACT	Coordinating the interrelated elements
38-ACT	Conducting design research
40-STRA	Considering critical path rather than critical chain
42-STRA	Having experienced designers conceive assumptions of similar designs
43-STRA	Having experienced designers prevent avoidable mistakes
45-ACT	Meeting collectively in conceptual phase to choose the mechanical and electronics concepts
46-ACT	Conducting initial test experiments in the preliminary design phase
47-ACT	Constructing to-scale cardboard prototype
48-ACT	Strength-adaptable mechanical chassis
50-ACT	Making sketches to generate concepts in the conceptual design phase
51-ACT	Making computer-aided design (CAD) models in the preliminary design phase
52-ACT	Building to-scale prototype to be tested in real operational environment

(Continued)

Table 6.3 (Continued) Uncorrelated design activities and strategies
with the technical attributes of product success

Design activities/strategies	
53-STRA	Considering manufacturability starting from the conceptual design phase
55-ACT	Determining optimal number of degrees of freedom of the system
56-STRA	Error tracing
57-ACT	Extending sensor's robustness through software manipulation
58-ACT	Minimising number of reads and writes and maximising utilisation of channeling and port addressing
59-ACT	Functional analysis and software architecture
61-ACT	Having time for overall testing
63-ACT	Implementing effective programming techniques
64-STRA	Implementing object-oriented programming
65-ACT	Implementing software code explanatory annotation
66-STRA	Implementing software code parameterisation

3. Moderate correlation when the correlation coefficient ranges from 0.3 to less than 0.6
4. High correlation when the correlation coefficient ranges from 0.6 to 1 (Cohen, 1988)

6.1.1 Positively correlated design activities and strategies with the technical attributes of product success

The positively correlated design activities and strategies to technical attributes of product success are presented in Table 6.1. The number that appears to the left of each design activity and strategy refers to its corresponding question in the relevant questionnaire as indicated in Appendices A, B and C. The design activities and strategies of positive correlation with technical attributes of product success should receive most focus and highest priority in resource allocation. This is determined according to their percentage of variation in technical performance due to the variable correlation coefficients with the technical performance attributes as a measure of success in the market, and according to their resulting p-value.

6.1.2 Negatively correlated design activities and strategies with the technical attributes of product success

The negatively correlated design activities and strategies with technical attributes of product success are presented in Table 6.2. The design activities and strategies of negative correlation with technical attributes of product success should receive the least focus and lowest priority in resource allocation.

6.1.3   Uncorrelated design activities and strategies with the technical attributes of product success

The design activities and strategies which are not correlated with technical attributes of product success are presented in Table 6.3. The design activities and strategies of no correlation with technical attributes of product success should receive low focus and low priority in resource allocation.

6.2   Results of analysis of bivariate correlation of the positively correlated design activity with the technical attributes of product success

The positively correlated design activity with the technical attributes of product success was to have the largest number of design iterations, if any, to occur within the software subsystem. This section shows frequencies and descriptive statistics of the positively correlated design activity with the technical attributes of product success. Table 6.4 shows that the total valid percentage of data was 99.4%, which is proof of valid results. Figure 6.1 depicts that the category of responses of 'Agree' had the largest percentage of responses, 42%, as to the design activity of having the largest number of iterations in the software subsystem. This section also presents nonparametric correlation analysis based on Spearman's ρ correlation. Table 6.5 shows that the design activity of

Table 6.4 Frequencies statistics of the design activity of having the largest number of iterations in the software subsystem

		Frequency	Percent	Valid percent	Cumulative percent
Valid	Strongly disagree	12	6.9	6.9	6.9
	Disagree	30	17.1	17.2	24.1
	Do not know	24	13.7	13.8	37.9
	Agree	72	41.1	41.4	79.3
	Strongly agree	36	20.6	20.7	100.0
	Total	174	99.4	100.0	
Missing	System	1	0.6		
Total		175	100.0		

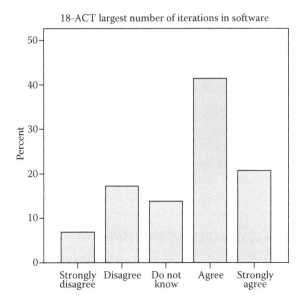

Figure 6.1 Percentage of responses in each category of response to having the largest number of iterations in software subsystem.

Table 6.5 Nonparametric correlations of the design activity of having the largest number of iterations in the software subsystem

			SCORE-NORMAL	18-ACT Largest number of iterations in software
Spearman's ρ	SCORE-NORMAL	Correlation coefficient	1.000	0.484(**)
		Significance (2-tailed)	–	0.000
		N	174	174
	18-ACT Largest number of iterations in software	Correlation coefficient	0.484(**)	1.000
		Significance (2-tailed)	0.000	–
		N	174	174

** Correlation is significant at the 0.01 level (2-tailed).

having the largest number of iterations in the software subsystem has a moderate correlation coefficient of 0.5 that is statistically significant at the 0.01 level of a 2-tailed analysis. In other words, this result is true to the population.

6.3   Results of analysis of bivariate correlation of the negatively correlated design activity with the technical attributes of product success

The negatively correlated design activity with the technical attributes of product success was to test aggregately the software subsystem at the end of the project rather than to adopt quick testing. This section shows frequencies and descriptive statistics of the negatively correlated design activity with the product success technical attributes. Table 6.6 shows that the total valid percentage of data was 99.4%, which is proof of valid results. Figure 6.2 depicts that the category of responses of 'Disagree' had the largest percentage of responses, 71%, as to the design activity of having the largest number of iterations in the software subsystem. This section also presents nonparametric correlation analysis based on Spearman's ρ correlation. Table 6.7 shows that the design activity of having the largest number of iterations in the software subsystem has a moderate correlation coefficient of 0.5 that is statistically significant at the 0.01 level of 2-tailed analysis. In other words, this result is true to the population.

Table 6.6 Frequencies statistics of the design activity of testing aggregately the software subsystem at the end of the project rather than adopting quick testing

		Frequency	Percent	Valid percent	Cumulative percent
Valid	Strongly disagree	30	17.1	17.2	17.2
	Disagree	126	72.0	72.4	89.7
	Agree	6	3.4	3.4	93.1
	Strongly agree	12	6.9	6.9	100.0
	Total	174	99.4	100.0	
Missing	System	1	0.6		
Total		175	100.0		

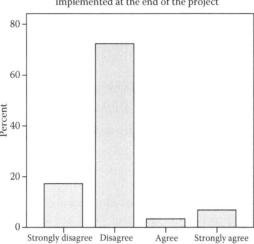

62-ACT adopting aggregate quota of testing time to be
implemented at the end of the project

Figure 6.2 Percentage of responses in each category of response to testing aggregately the software subsystem at the end of the project rather than adopting quick testing.

Table 6.7 Nonparametric correlations of the design activity of testing aggregately the software subsystem at the end of the project rather than adopting quick testing

			SCORE-NORMAL	62-ACT Testing aggregately the software subsystem at the end of the project rather than adopting quick testing
Spearman's ρ	SCORE-NORMAL	Correlation coefficient	1.000	−0.507(**)
		Significance (2-tailed)	–	0.000
		N	174	174
	62-ACT Testing aggregately the software subsystem at the end of the project rather than adopting quick testing	Correlation coefficient	−0.507(**)	1.000
		Significance (2-tailed)	0.000	–
		N	174	174

** Correlation is significant at the 0.01 level (2-tailed).

6.4 Results of dependency analysis using bivariate correlation and partial correlation analyses

To test the rigour of the bivariate correlation analysis results of Sections 6.2 and 6.3, a dependency analysis is conducted. The dependency analysis explores whether or not an independent variable, which was proven to be correlated with a dependent variable, is in turn a dependent variable on other variables. The dependency analysis adopted in this book is twofold. First, a mutual dependency analysis based on a bivariate correlation coefficient is conducted. Second, partial correlation analysis is conducted consequently to control for the effect of each of the two mutually dependent variables on each other in relation to other variables.

In this work, the only pair of design activities and strategies that has mutual dependency was design strategy 6, which is to adopt modular design, and design strategy 11, which is to strike a balance between functionality and design iterations. The bivariate correlation coefficient between these two design strategies was 0.71, as shown in Table 6.8, which implies a strong potential for mutual dependency. Hence, a partial correlation analysis between strategy 6 and product performance attributes, controlling for strategy 11, was conducted as shown in Table 6.9.

The result showed that the effect of strategy 11 on the relationship between strategy 6 and the product performance attributes is negligible because it slightly changed the correlation coefficient within the same order of magnitude. Consequently, the correlation coefficient between strategy 6 and the product performance attributes will remain unchanged in the low correlation category. Also, a partial correlation analysis between strategy 11 and product performance attributes, controlling for strategy 6, was conducted as shown in Table 6.10.

The result showed that the effect of strategy 6 on the relationship between strategy 11 and the product performance attributes is negligible because it slightly changed the correlation coefficient within the same order of magnitude. Consequently, the correlation coefficient between strategy 11 and the product performance attributes will remain unchanged in the no correlation category.

Table 6.8 Result of dependency analysis – bivariate correlation

			6-STRA Adopting a modular design
Spearman's ρ	11-STRA Striking a balance between functionality and design iterations	Correlation coefficient	−0.711
		Significance (2-tailed)	0.000
		N	266

Table 6.9 Partial correlation between strategy 6 and normalised performance score, controlling for strategy 11

Control variables			6-STRA Adopting a modular design	SCORE-NORMAL
11-STRA Striking a balance between functionality and design iterations	6-STRA Adopting a modular design	Correlation	1.000	0.049
		Significance (2-tailed)	–	0.427
		Df	0	263
	SCORE-NORMAL	Correlation	0.049	1.000
		Significance (2-tailed)	0.427	–
		Df	263	0

Table 6.10 Partial correlation between strategy 11 and normalised performance score, controlling for strategy 6

Control variables			SCORE-NORMAL	11-STRA Striking a balance between functionality and design iterations
6-STRA Adopting modular design	SCORE-NORMAL	Correlation	1.000	0.009
		Significance (2-tailed)	–	0.890
		Df	0	263
	11-STRA Striking a balance between functionality and design iterations	Correlation	0.009	1.000
		Significance (2-tailed)	0.890	–
		Df	263	0

6.5   Reliability analysis results

In reliability statistics, if the data collected achieved 0.7 or more on the Cronbach's α scale, the collected data would then have good internal consistency (Sekaran, 2003). Because the Cronbach's Alpha internal reliability factor of data collected is 0.7, as shown in Table 6.11, there is good internal consistency, based on the average interitem correlation, of the data collected.

Table 6.11 Reliability statistics

Cronbach's α	Cronbach's α based on standardised items	N of items
0.678	0.705	67

6.6   Assumptions in the reliability analysis

The reliability analysis was conducted based on some assumptions.

1. Observations should be independent.
2. Errors should be uncorrelated between items, that is, between the design activities and strategies.

6.7   Verification and validation of the results of the quasi-experiments

As to the research verification that should answer the question 'Did we measure the thing right?' this book adopted measuring the Spearman correlation coefficient to examine the hypothesised correlation between the rank of adoption of design activities and strategies by the designers and the technical attributes of the performance of their design outcomes and products. In addition, the data were collected according to standard statistical methodology. Therefore, this book adopts a nonparametric statistical analysis to avoid making any assumption about the population parameters and is expected to provide more rigourous results than the parametric statistical analysis. In addition, the reliability statistics test results further proved the reliability of the data and consequently verification of the results.

As to the research validation, an inferential statistical analysis was conducted on the resulting correlation coefficients. The p-value was adopted as a measure that a result is true to the population. A cutoff p-value of 0.1 is adopted in this book. Causality in this work is determined according to the percentage of variation in technical performance due to the variable (r^2) of moderate-to-high correlation coefficient (r). The research results were validated successfully in terms of the four validity types: first, in terms of statistical conclusion validity, because the resulted relationships were meaningful and reasonable; second, in terms of internal validity, because the results were causal rather than just descriptive; third, in terms of construct validity, because the results represented what is theoretically intended and fourth, in terms of external validity, because the results could be generalised to the population.

The statistical sampling in this book was representative in terms of sampling design that was suitable for generalisation with cost-effectively fair statistical results and sample size that satisfied the minimum statistically representative sample size. In addition, the data used in this study satisfied the three aspects of a statistical power test: first, the statistical significance aspect was satisfied in terms of proved significance of research results at the 2-tailed 0.01 level; second, the size of the population was reasonably satisfied because despite a large population size, which can negatively affect the statistical power of the data, a probability sampling design was used to mitigate such a negative effect; third, the reliability of the data was fairly satisfied through passing the data reliability and verification test as indicated in this chapter.

6.8   Research validation and verification of the hybrid approach through case studies

In an endeavor to investigate further the validation and verification of the hybrid approach, a case study was conducted. In this case study, three companies were studied with respect to each of the aforementioned approaches. The descriptive characteristics of these companies are presented in Table 6.12 (Bigliardi and Galati, 2014).

The level of adoption that each of these three companies have reached with respect to each of the aforementioned approaches is presented in Tables 6.13 to 6.17. The scale of evaluation on each practice within each of

Table 6.12 Descriptive characteristics of the three companies

Company	Year of establishment	Industry	Number of employees	Turnover per year
Company I	1994	Food machinery	40	<2 million euro
Company II	1976	Mechanical	143	10–25 million euro
Company III	1997	Pharmaceutical	146	<2 million euro

Table 6.13 The level of adoption of the strategic, resources-based and sustainability-oriented approach

Aspect	Practice	Company I	Company II	Company III
Strategic, resources-based and sustainability-oriented aspect	Setting and implementing R & D Strategy	3.8	3.3	3
	Early market entrants	4	5	3
	Mean	3.9	4.2	3

Table 6.14 The level of adoption of the cost-oriented approach

Aspect	Practice	Company I	Company II	Company III
Cost-oriented aspect	Employee flexibility and multiskilling	5	5	4
	Financial performance	4.2	4	2
	Mean	4.6	4.5	3

Table 6.15 The level of adoption of the time duration-oriented approach

Aspect	Practice	Company I	Company II	Company III
Time duration-oriented approach	Speed of product development	4	4	2
	Number of new products	5	5	2
	Mean	4.5	4.5	2

Table 6.16 The level of adoption of the technical and problem solving approach

Aspect	Practice	Company I	Company II	Company III
Technical and problem solving aspect	Partnering with suppliers to identify needs and requirements	5	5	3
	Performance of products	5	5	3
	Mean	5	5	3

Table 6.17 The level of adoption of the social science and knowledge-based approach

Aspect	Practice	Company I	Company II	Company III
Social science and knowledge-based aspect	Up-to-date data and information is readily available	4	4	3
	Maintain a close relationship with our customers	5	4	2
	Mean	4.5	4	2.5

the aforementioned approaches is a five-point scale; that is, level 5 is the highest level. Tables 6.13 to 6.17 present these levels of adoption.

Tables 6.13 to 6.17 indicate that companies I and II are much more successful than company III. Tables 6.13 to 6.17 indicate that only the companies that have adopted the hybrid approach have succeeded in all of the five aspects: (1) strategic, resources-based and sustainability-oriented aspect; (2) cost-oriented aspect; (3) time duration-oriented aspect; (4) technical and problem solving aspect and (5) social science and knowledge-based aspect.

6.9   Implication of percentage of variation in technical performance due to the design variables (r^2)

The aggregation of percentage of variation in technical performance due to the design variables reached collectively more than 100% because there is overlap in the affected areas in the product technical performance by the design variables. The design variables have been proven as independent except for 3% of them, where mutual dependency was identified as proved in the dependency analysis results in Section 6.4.

chapter seven

Implications and conclusions of research findings

This chapter presents the implications and conclusions of the findings of the quasi-experiments investigated. It sorts out the moderately correlated and uncorrelated design activities and strategies with product success technical attributes classified according to the dimensions presented. The implications of the experimental observations and research hypotheses are discussed.

7.1 Activities and strategies of the system design dimension that are correlated and uncorrelated with the technical attributes of product success

System design was the first of the five dimensions of the integrated design process. It addresses the interrelationship between the various parts of the process and/or of the product and aims to make the whole greater than the sum of its parts. In this work, it was found that there are eight design activities and strategies within the system design dimension that are moderately positively correlated with the technical attributes of product success (Table 7.1). Such design activities and strategies should receive the most focus and highest priority in resource allocation.

In addition, there are five activities and strategies of the system design dimension that have no correlation with the technical attributes of product success (Table 7.2). Such design activities and strategies should receive low focus and low priority in resource allocation.

Moreover, it was found that there are no activities and strategies of the system design dimension that are used by designers in an integrated design project and negatively have moderate correlation with the technical attributes of product success.

Table 7.1 System design dimension activities and strategies that are positively correlated with the technical attributes of product success

System design hypothesis/observation	Design phase	Design strategy/ activity	Hypothesis/ observation
1. In the integrated design system, shifting complexity towards the software subsystem, rather than the mechanical subsystem, to have the largest number of design iterations, if any, to occur within the software subsystem, followed accordingly by the number of iterations in the electronics subsystem	Preliminary and detailed design phases	Activity	Observation 10
2. In the integrated design system, having iterations in the software subsystem, rather than in the mechanical subsystem, achieves a shorter development time	Preliminary and detailed design phases	Activity	Observation 2
3. In the integrated design system, considering reliability of the final product in the design process, in terms of the ability of the product to perform its required functions under stated conditions for a specified product service time	Conceptual and preliminary design phases	Strategy	Hypothesis 8
4. In the integrated design system, having designs that are less vulnerable to failure modes (e.g. out-of-plane buckling, roll mode and pitch mode instability, and friction wear) (and consequently have better reliability) and are less exposed to and less sensitive to the uncontrollable external factors (and consequently had better robustness), by shifting complexity to the software subsystem rather than to the mechanical subsystem	Conceptual and preliminary design phases	Strategy	Observation 1

(Continued)

Table 7.1 (Continued) System design dimension activities and strategies that are positively correlated with the technical attributes of product success

System design hypothesis/observation	Design phase	Design strategy/ activity	Hypothesis/ observation
5. In the integrated design system, having fewer constraints on the design by shifting complexity to the software subsystem and consequently to the virtual space, for example, data processing time, rather than to the mechanical subsystem and consequently to the physical space, for example, suspension space	Conceptual and preliminary design phases	Strategy	Observation 3
6. In the integrated design system, aiming at striking a balance between fast response on one hand and stability, accuracy and payload fulfilment of the final concept of the product on the other	Conceptual and preliminary design phases	Strategy	Hypothesis 7
7. In the integrated design system, adopting top-down system structure decomposition to analyse the functional structure of the product and consequently to map it to the requirements specification and consequently come up with the product functional requirements	Scoping phase	Activity	Hypothesis 6
8. Adopting the simplest design, which addresses the design problem requirements specification by the minimum set and most effective combination of system components, rather than the cheapest and the lightest weight design	Conceptual and preliminary design phases	Strategy	Observation 4

Table 7.2 System design dimension activities and strategies that are uncorrelated with the technical attributes of product success

System design hypothesis/observation	Design phase	Design strategy/ activity	Hypothesis/ observation
1. In the integrated design system, considering safety of use of the final product in the design process, in terms of accident prevention, risks identification and risk control and/or minimisation	Conceptual and preliminary design phases	Strategy	Hypothesis 10
2. In the integrated design system, adopting a systematic design approach, that is, (1) clarifying the design problem; (2) identifying requirements specification and design constraints; (3) sorting the requirements specification in a ranked order; (4) identifying functional requirements, for example, using an axiomatic design methodology, and identifying core design values; (5) constructing a table of design options based on brainstorming; (6) making concept evaluation, for example, based on a weighted criteria evaluation matrix and (7) making an overall strategic decision of the conceptual design	Scoping and conceptual design phases	Activity	Hypothesis 2
3. In the integrated design system and in case of adopting the systematic design approach, paying special attention to the quality of performing each of the steps of the systematic design approach	Scoping and conceptual design phases	Activity	Hypothesis 2
4. In the integrated design system, setting an error-proof operational design strategy, for example, constructing a 3D computer-aided design (CAD) model of the system assembly drawing to minimise waste of time and resources that is due to making avoidable mistakes	Preliminary and detailed design phases	Strategy	Hypothesis 4
5. In the integrated design system, aiming at striking a balance between maximising functionality of the product final concept on one hand and minimising waste of time and resources by minimising design iterations on the other	Conceptual and preliminary design phases	Strategy	Hypothesis 5

7.2 Design activities and strategies of the project management dimension that are correlated and uncorrelated with the technical attributes of product success

As the second of the five dimensions of the integrated design process, project management reflects the interaction among human resources, facilities, budget and strategic goals. In this work, it was found that there are 10 design activities and strategies within the project management dimension that have a moderately positive correlation with the technical attributes of product success (Table 7.3). Such design activities and strategies should receive the most focus and highest priority in resource allocation.

In addition, there are 12 activities and strategies of the project management dimension that have no correlation with the technical attributes of product success (Table 7.4). Such design activities and strategies should receive a low focus and low priority in resource allocation.

7.3 Design activities and strategies of the mechanical subsystem dimension that are correlated and uncorrelated with the technical attributes of product success

The mechanical design subsystem was the third of the five dimensions of the integrated design process. The mechanical design subsystem relates to the skeleton of the product that deals with physical external loads. In this work, it was found that there are two design activities and strategies within the mechanical design subsystem dimension that have a positive correlation with the technical attributes of product success (Table 7.5). Such design activities and strategies should receive the most focus and highest priority in resource allocation.

In addition, there are three activities and strategies in the mechanical design subsystem dimension that have no correlation with the technical attributes of product success (Table 7.6). Such design activities and strategies should receive a low focus and low priority in resource allocation.

Moreover, it was found that there are no activities and strategies of the mechanical design subsystem dimension that are used by designers in an integrated design project and have a moderately negative correlation with the technical attributes of product success.

Table 7.3 Project management dimension activities and strategies that are moderately positively correlated with the technical attributes of product success

Project management hypothesis/observation	Design phase	Design strategy/ activity	Hypothesis/ observation
1. In the integrated design system, adopting testable design interdeliverables within and among system modules based on project milestones, to detect mistakes as early as possible and to minimise their impact on the successful completion of the design project	Preliminary and detailed design phases	Activity	Hypothesis 17
2. In the integrated design system, adopting early verification and validation of the design concept, for example, early testing in the design process, to avoid becoming trapped in incompetent design concepts	Conceptual and preliminary design phases	Activity	Observation 11
3. In the integrated design system, given the intertwined and overlapping nature of the subsystems of the product, for example, the mechanical–electronics interconnection for robot speed, adopting modular deliverables and testing, that is, testing deliverables of each module, rather than subsystems deliverables and testing	Preliminary and detailed design phases	Activity	Hypothesis 23
4. Having resource provisions for unforeseen problems, to minimise vulnerability of our design development process to the influence of external factors	Scoping phase	Strategy	Hypothesis 15
5. In the integrated design system, having a multidisciplinary team of designers, each of whom is aware of more than one of relevant disciplines such as material science, design development approaches, leadership skills, and so forth, and is aware of the overlap and intersections between them, to improve the opportunity for achieving a better design and to minimise the risk of mistakes	Scoping phase	Activity	Hypothesis 14

(Continued)

Table 7.3 (Continued) Project management dimension activities and strategies that are moderately positively correlated with the technical attributes of product success

Project management hypothesis/observation	Design phase	Design strategy/ activity	Hypothesis/ observation
6. In the integrated design system, setting an operational design strategy of modular testing, that is, testing the deliverables between the system submodules as a way of verifying conformance of these system submodules to the conceptual functional requirements, to detect mistakes as early as possible and to minimise their impact on the successful completion of the design project	Conceptual and preliminary design phases	Strategy	Hypothesis 16
7. In the integrated design system, adopting a decentralised decision-making strategy, by empowering and authorising subteams to make tactical decisions without need to refer them to the team leader, rather than strategic decisions that can degrade performance in this case	Conceptual, preliminary and detailed design phases	Strategy	Hypothesis 19
8. In the integrated design system, documenting the outcome of design discussions and consequently analysing that outcome	Conceptual and preliminary design phases	Activity	Hypothesis 26
9. In the integrated design system, meeting collectively in the conceptual design phase than in detailed design phase	Conceptual and detailed design phases	Activity	Hypothesis 29
10. In the integrated design system, starting the design process as early as possible in the project timeframe and assigning sufficient time for presenting the outcome of the design process milestones	Scoping and conceptual design phases	Activity	Observation 18

Table 7.4 Activities and strategies of the project management dimension that are uncorrelated with the technical attributes of product success

Project management hypothesis/observation	Design phase	Design strategy/activity	Hypothesis/observation
1. Adopting cross fertilisation of design concepts among the design team, such as sharing ideas and/or modifying each other's ideas	Conceptual and preliminary design phases	Activity	Hypothesis 13
2. In the integrated design system, adopting live meetings of the design team in the conceptual and preliminary design phases, and e-mails in the detailed design phase, as the principal communication method, to strike a balance between the synergy of face-to-face meetings on one hand and the time effectiveness and message delivery effectiveness of the electronic means of communication, such as e-mails, on the other	Conceptual, preliminary and detailed design phases	Activity	Hypothesis 18
3. Adopting both sequential and concurrent design activities according to the nature of each task and the preceding relationships between them, to maximise utilisation of resources and shorten the design development time	Scoping phase	Strategy	Hypothesis 20
4. In the integrated design system, being more interested in meeting project delivery deadlines than in achieving the project objectives to a specified cost	Scoping and conceptual and preliminary and detailed design phases	Strategy	Hypothesis 21

(*Continued*)

Table 7.4 (Continued) Activities and strategies of the project management dimension that are uncorrelated with the technical attributes of product success

Project management hypothesis/observation	Design phase	Design strategy/activity	Hypothesis/observation
5. In the integrated design system, adopting redeployment of the human resources of the design team to cope with fluctuations in the workload of project activities	Conceptual, preliminary and detailed design phases	Strategy	Hypothesis 22
6. In the integrated design system, conducting a design research activity (e.g. research literature review, review of previous student reports, etc.) and/or Internet literature review search to become acquainted with what other designers have done to design and implement similar products and to be acquainted with the state of the art of the design of the required product	Scoping phase	Activity	Hypothesis 24
7. In the integrated design system, considering the total time duration of the project as the most important constraint and consequently being more interested in the critical path, the longest sequence of time-constrained design tasks in the project plan, than in the critical chain, the longest sequence of resource-constrained design tasks in the project plan	Scoping phase	Strategy	Hypothesis 25
8. In the integrated design system, having experienced designers in the design team, to better conceive the assumptions and constraints accompanied by similar designs	Scoping phase	Strategy	Hypothesis 27

(Continued)

Table 7.4 (Continued) Activities and strategies of the project management dimension that are uncorrelated with the technical attributes of product success

Project management hypothesis/observation	Design phase	Design strategy/activity	Hypothesis/observation
9. In the integrated design system, having experienced designers in the design team, to have a foolproof design process and to prevent avoidable mistakes in the design process, such as mounting the more heat-sensitive electronic components first on the printed circuit board (PCB) prototype	Scoping phase	Strategy	Hypothesis 28
10. In the integrated design system, selecting team leader with a software specialisation and familiarity with electronics and mechanical subsystems	Scoping phase	Activity	Observation 12
11. In the integrated design system, estimating the duration of each project activity based on work breakdown structure, experience and comparison with comparable projects	Scoping phase	Activity	Observation 16
12. In the integrated design system, coordinating between the design subteams concerning the interrelated elements of the system, for example, software–electronics interconnection for microchip pin allocation and for positioning of sensors that fits all the required routing	Preliminary and detailed design phases	Activity	Observation 17

Table 7.5 Positively correlated mechanical design subsystem dimension activities and strategies with product success technical attributes

Mechanical design subsystem hypothesis/ observation	Design phase	Design strategy/ activity	Hypothesis/ observation
1. Checking accuracy of manufacturing and assembly of the final prototype, to avoid unexpected failure due to manufacturing defects and/or assembly mistakes	Detailed design phase	Activity	Hypothesis 31
2. Making a prediction of the progressive failure of the design through drawing sketches or through conducting a finite element analysis, to avoid unexpected failure	Conceptual and preliminary design phases	Activity	Observation 23

7.4   Activities and strategies of the electronics subsystem dimension that are correlated and uncorrelated with the technical attributes of product success

The aim of the electronics design subteam is to design, build and test the interface electronic circuits that will be the interface between the interdisciplinary system's transducers, which provide information on the 'real world', and the interdisciplinary system's microprocessor.

In this work, it was found that there is an activity of the electronics design subsystem dimension that has no correlation with the technical attributes of product success (Table 7.7). Such a design activity should receive a low focus and low priority in resource allocation.

Moreover, it was found that there are no activities and strategies of the electronics design subsystem dimension that are used by designers in an integrated design project and have either moderately positive or negative correlation with the technical attributes of product success.

Table 7.6 Activities and strategies of the mechanical design subsystem dimension that are uncorrelated with the technical attributes of product success

Mechanical design subsystem hypothesis/ observation	Design phase	Design strategy/ activity	Hypothesis/ observation
1. Conducting a to-scale cardboard prototype, mock-up, or equivalent in the mechanical subsystem	Preliminary design phase	Activity	Observation 21
2. Building a to-scale design prototype to be tested in the real operational environment in the detailed design phase	Detailed design phase	Activity	Observation 26
3. In the integrated design system, determination by the mechanical subteam of the optimal number of degrees of freedom of the system, for example, of picking-up and unloading mechanisms, that fulfil the functional requirements and minimise the number of needed controllable mechanisms	Preliminary and detailed design phases	Activity	Observation 27

Table 7.7 Activities and strategies of the electronics design subsystem dimension that are uncorrelated with the technical attributes of product success

Electronics design subsystem hypothesis/ observation	Design phase	Design strategy/ activity	Hypothesis/ observation
In the integrated design system, developing the electronics circuitry such that it minimises the number of required reads and writes, and maximises utilisation of channelling and port addressing	Conceptual, preliminary and detailed design phases	Activity	Observation 30

7.5 Design activities and strategies of the software subsystem dimension that are correlated and uncorrelated with the technical attributes of product success

The fifth dimension of the integrated design process was the software design subsystem. The software design subsystem applies software architecture design and program module code design techniques and procedures to the development, operation and maintenance of software. In this work, it was found that there is a design activity within the software design subsystem that has a positive correlation with the technical attributes of product success (Table 7.8). Such a design activity should receive the most focus and highest priority in resource allocation.

In addition, it was found that there is an activity in the software design subsystem dimension that has a moderately negative correlation with the technical attributes of product success (Table 7.9). Such a design activity should be eliminated from the design process.

Moreover, there are three activities and strategies of the software design subsystem dimension that have no correlation with the technical attributes of product success (Table 7.10). Such design activities and strategies should receive a low focus and low priority in resource allocation.

Table 7.8 Activities and strategies of the software design subsystem dimension that are positively correlated with the technical attributes of product success

Software design hypothesis/ observation	Design phase	Design strategy/ activity	Hypothesis/ observation
In the integrated design system, adopting quick testing of interdeliverables between the modules of the software draft code and extensive testing of the overall software draft code on a prototype PCB or on an equivalent facility, to strike a balance between minimising cost of test and detecting mistakes as early as possible	Preliminary and detailed design phases	Activity	Hypothesis 33

Table 7.9 Activities and strategies of the software design subsystem dimension that are negatively correlated with the technical attributes of product success

Software design subsystem hypothesis/observation	Design phase	Design strategy/ activity	Hypothesis/ observation
In the integrated design system, testing aggregately the software subsystem at the end of the project rather than adopting quick testing	Scoping phase	Activity	Observation 33

Table 7.10 Activities and strategies of the software design subsystem dimension that are uncorrelated with the technical attributes of product success

Software design subsystem hypothesis/observation	Design phase	Design strategy/ activity	Hypothesis/ observation
1. In the integrated design system, aiming at implementing explanatory annotation of the software code for purposes of software code debugging and maintenance	Preliminary and detailed design phases	Activity	Hypothesis 35
2. In the integrated design system, adopting a parameterisation design strategy, that is, changing the value of one parameter changes the values of many related parameters accordingly, to facilitate software code debugging, maintainability and/or scalability of the final concept of the product	Preliminary and detailed design phases	Strategy	Hypothesis 36
3. In the integrated design system, spending at least 10% of the total time duration of the software subsystem on functional analysis and software architecture, for example, making program flow chart, rather than on the programming code implementation itself, to minimise waste of time and resources and to eliminate sources of avoidable mistakes	Scoping phase	Activity	Observation 31

7.6 Implications of the experimental observations

The objective of this book is to set engineering design guidelines in terms of engineering design activities and strategies that can help designers and design managers to manage the design process more cost-effectively. In this book, three types of studies were adopted:

1. 'Exploratory study' to understand the design process and to define research questions and identify hypotheses
2. 'Descriptive study' to describe the collected data, for example, in terms of the mean and standard deviation
3. 'Correlational study' to explore correlation and causality

The following implications are drawn from the experimental observations in this work.

1. According to Observation 4 in Table 7.1, adopting the simplest design and determining the minimum set and most effective combination of system components should be among the most important design activities.
2. Shifting complexity towards the software subsystem, rather than the mechanical subsystem, such as in Observation 10 in Table 7.1, leads to reducing time and cost of testing the product prototype, facilitates scalability of the product and reduces the cost of adaptability.
3. Shifting complexity from the mechanical towards the software subsystem, according to Observation 1 in Table 7.1, should be accompanied by building electronics as soon as possible, to leave enough time for software testing.
4. Design robustness, on the basis of Observations 1, 7, 9 and 29 in Table 5.1, has three dimensions: (1) enabling the robot to deal with situations of abnormal operating conditions, (2) limiting exposure to uncontrollable external factors and (3) decreasing sensitivity to uncontrollable external factors.
5. Almost none of the respondents adopted all of the hypothesised and observed design activities and strategies.
6. Adopting only one or a few of the most important design activities and strategies did not guarantee success of the product; therefore, the aim of this work is to introduce a design guideline of design activities and strategies to be adopted together with a level of focus and resources allocation that is proportional to the level of variation in the product's technical performance as a result of those design activities and strategies. For instance, aiming at a simple design, as mentioned in Observation 4 in Table 5.1, can be a recommended strategy; however, focussing only on this target and exaggerating

the simplicity of the design can have a negative impact, as a balance should be struck among the design strategies as well as among the design activities. Therefore, this proposed design guideline helps to identify this balance.

7. There could be two justifications for Observation 12 in Table 5.1, often found in many design teams, that is, selecting a team leader with a software specialisation and familiarity with electronics and mechanical subsystems. First, the software subsystem is intangible and difficult to be conceived by unspecialised designers. Second, software subsystem design effectively starts after the mechanical and electronics design concepts are decided on and consequently the software designer has a better opportunity to see and manage the overall project.

7.7   Implications of the research hypotheses

The research hypotheses proposed and verified in this work raise some implications, as follows:

1. Regarding step 5 of Hypothesis 2 in Table 4.1, by constructing a table of design options based on brainstorming, designers would cross fertilise each other's design ideas to achieve viable design options.
2. Regarding step 6 of Hypothesis 2 in Table 4.1, making a concept evaluation, for example, based on a weighted criteria evaluation matrix, has the following implications:
   a. This is the most subjective step in that approach; therefore, each design concept evaluation has to be justified.
   b. The person who judges the design concept has to be experienced in design with relevant multidisciplinary backgrounds.
   c. More than one judge should evaluate each design individually and then meet collectively to discuss and choose the most convincing weighted factor for each design concept.
3. Regarding Hypothesis 6 in Table 4.1, adopting top-down system structure decomposition to analyse and understand the functional structure of the product, in larger and similar design projects can help to understand better how design changes propagate bottom-up and laterally.
4. Some design activities and strategies can have hidden negative consequences in contrast with what is usually perceived about those design activities and strategies. For instance, according to Hypothesis 12 in Table 4.1, visiting and reviewing similar designs that solved similar design problems are usually perceived to improve the designers' opportunities to conceive and understand what the

functional units of that product are, what those functional units can look like and how they work. Consequently, it is usually perceived that this improves the designers' opportunities of achieving a better design. However, this design activity can have a negative consequence; that is, it relatively confines the designer's mental images in the brainstorming activity in the conceptual design phase within the borders of the designs that the designer visited and reviewed previously. For example, in the Integrated Design Project, designers were given a dummy chassis for initial experimental design trials until each design team designed and built its own mechanical design concept and chassis. It was noted that a large percentage of the students opted for mimicking the dummy chassis that they were given. Therefore, it is recommended that the designer who visits and reviews similar designs that solved similar design problems should not be a member of the brainstorming subteam but instead a member of the design concept evaluation subteam.

7.8 Implications of the main findings of the statistical analysis

The design activities and strategies that are positively correlated with the technical attributes of product success, which should receive the most focus and highest priority in resource allocation, fit with the literature. Such design activities and strategies have a significantly positive influence on variations in the technical attributes of product success – more than 10%. The design activities and strategies that are uncorrelated with the technical attributes of product success, which should receive a low focus and low priority in resource allocation, are interpreted in this work as design activities and strategies that are adopted by all designers as they are important to success but are not success differentiators. Their importance to success is either proven and supported by the literature or intuitively perceived by the engineering designers.

In the system design dimension, there were unexpectedly three design activities and strategies that were uncorrelated with the technical attributes of product success although their importance to success is supported by the literature. These three design activities and strategies are Hypothesis 2, which is to adopt a systematic design approach; Hypothesis 4, which is to set an error-proof operational design strategy and Hypothesis 5, which is to strike a balance between maximising product's functionality and minimising iterations. The justification for this is that because they are backed by literature, they have become widely adopted by engineering designers and consequently they have not been success differentiators. In the project management dimension, there were unexpectedly two design activities and strategies that were uncorrelated with the technical attributes product success although their importance to success is supported by the literature. These two design activities and strategies are Hypothesis 13, which is to adopt cross fertilisation of design concepts among the design team, and Hypothesis 20, which is to adopt both sequential and concurrent design activities according to the nature of each task and the preceding relationship between them. The justification for this is that because they are backed by literature, they have become widely adopted by the engineering designers and consequently have not been success differentiators. In the mechanical design subsystem dimension, there was unexpectedly a design activity that was uncorrelated with the technical attributes of product success although its importance to success is intuitively perceived. This design activity is Observation 26, which is to build a to-scale design prototype to be tested in the real operational environment. In the electronics design subsystem dimension, there was unexpectedly a design activity that was uncorrelated with the technical attributes of product success although its importance to success is intuitively perceived. This design activity is Observation 30, which is to minimise the number of reads and writes and to maximise utilisation of channelling and port addressing. The justification for this is also that because their importance to success is intuitively perceived by engineering designers, they have become widely adopted and consequently they have not been success differentiators.

The three design activities and strategies identified as having a moderately negative correlation with the technical attributes of product success should be removed from the design process. Such design activities and strategies have a significantly negative influence on variations in the technical attributes of product success – more than 10%. The first of these was the project management dimension's Observation 13, as noted in Table 5.1, which is to spend more time on the conceptual and preliminary design phases together than on the detailed design phase. Although it was unexpected to find that this design activity is negatively correlated with the technical attributes of the product success, there are three justifications for this.

1. There is a cost and time effectiveness of having all the planned iterations take place in the conceptual and preliminary design phases.
2. It addresses the best practice of having no more than five design concepts in the conceptual design phase to avoid distracting designers' focus.
3. It addresses the concept of Acceptable Quality Level that cost and time effectively satisfies the functional requirements.

Therefore, it is better to have a shorter time for conceptual and preliminary design phases than for the detailed design and testing time.

The second of the activities was the project management dimension's Observation 20, which is to conduct initial test experiments in the preliminary design phase. Such a design activity should be removed from the design process because it can be substituted by having experienced designers in the design process, which has proved helpful to success in the literature as mentioned in Hypotheses 27 and 28 in Table 4.1, and is proved to be positively correlated with the technical attributes of product success as well.

The third of the activities was the software design subsystem dimension's Observation 33, as mentioned in Table 5.1, which is to have the overall test time at the end of the project rather than to adopt early quick testing. Such a design activity should be removed from the design process and the overall test time should not be concentrated at the end of the project because quick testing proved to be more cost effective and time effective; for instance, quick testing costs 133% less in comparison with the cost of detailed testing and is more appropriate for the proven successful incremental innovation approach as supported by the literature, as noted in Hypothesis 33 in Table 4.1. In addition, quick testing proved to have a moderately positive correlation with product success technical attributes. Therefore, both quick testing throughout the conceptual and preliminary design phases and detailed testing at the end of the detailed design phases should be adopted with overall test time of at least 19% of the overall project time, as proved statistically.

Interestingly, it was unexpectedly found by Observation 31 in Table 5.1 that spending a large percentage of total time in the software subsystem on functional analysis and software architecture is uncorrelated with the technical attributes of product success. Therefore, it is recommended to spend no more than 12% of total time in the software subsystem on functional analysis and software architecture, as proved statistically. Also, it was found from Observation 14 in Table 5.1 that making two plans, a macroscale aggregate plan in the scoping phase and a detailed design plan at the preliminary design phase, has a low negative correlation with the technical attributes of product success. Therefore, it is recommended either to make one aggregate plan in the scoping phase of the design process and to have resource provisions, for example, time buffer, for variations in details, or to spend no more than 5% of the project time on making a macroscale aggregate plan in the scoping phase and a detailed design plan at the preliminary design phase, as proved statistically.

The other interesting finding was related to modularity in engineering design. Although design modularity is supported in the literature as important to design success, adopting a modular design strategy, which is to split the overall system into modules that interface with each other as mentioned in Hypothesis 1 in Table 4.1, had a low positive correlation with the technical attributes of product success. However, the other three design modularity-related activities and strategies, Hypothesis 23 from Table 4.1, which is to adopt modular deliverables and testing rather than subsystem deliverables and systems; Hypothesis 16 from Table 4.1, which is to adopt modular testing as an early testing approach by testing deliverables between system submodules against the conceptual functional requirements and Hypothesis 33 from Table 4.1, which is to adopt both quick testing of interdeliverables between submodules and overall testing of the system, proved to have a moderately positive correlation with the technical attributes of product success. Therefore, aggregately on average the four design modularity-related design activities and strategies have a moderate correlation with the technical attributes of product success. This implies that adopting a modular design strategy by splitting the overall system into modules that interface with each other is not sufficient to achieve a technically successful modular design. It should be accompanied by three other related and complementary design activities and strategies to achieve technically successful modular designs: adopting modular deliverables and testing rather than subsystem deliverables and systems, adopting modular testing as an early testing approach and adopting both quick testing of interdeliverables between submodules and overall testing of the system.

7.9 Hybrid lean–agile design paradigm

The design activities and strategies that correlated positively with product success can be classified into lean design activities and strategies and agile design activities and strategies (Elmoselhy, 2015) (Tables 7.11 and 7.12).

A key reason to have more lean than agile design strategies/activities is that the lean design paradigm is basically more related to the key determinants of the mobile robot development process indicated in Figure I.1 in the Introduction than the agile design paradigm. In addition, lean design strategies are basically more intuitive to designers than agile design strategies. Key challenges in the hybrid lean–agile mobile robot design paradigm include cost-effectively striking a balance between robot quality and short duration of the design process. In addition, misreporting cost savings can hurt the credibility of the hybrid lean–agile mobile robot design paradigm practitioners. Moreover, lean savings are indeed a long-term proposition. There is expectedly much pressure imposed on the implementers of the hybrid lean–agile mobile robot design paradigm to show immediate savings in terms of financial indicators such as the Return on Investment (ROI) to top management. Yet, reduction in defects and reduced cycle times are all areas that will continue to produce savings long after the term of ROI has run out. The implementation of the hybrid lean–agile mobile robot design paradigm thus needs a champion to lead cost-effectively the change and leverage it.

Table 7.11 Lean activities and strategies that are positively correlated with mobile robot performance

Observation/hypothesis	Design phase	Design strategy/ activity	Observation/ hypothesis	Reference in literature to hypothesis	Percentage of variation in mobile robot performance (r^2)
14-STRA Considering reliability of the mobile robot in the design process, in terms of the ability of the mobile robot to perform its required functions under stated conditions for a specified robot service time	Conceptual and preliminary design phases	Strategy	Hypothesis	Coulibaly et al., 2008; Prabhakar Murthy et al., 2008	0.2**
29-ACT Adopting testable design interdeliverables within and among system modules based on project milestones, to detect mistakes as early as possible and to minimise their impact on the successful completion of the design project	Preliminary and detailed design phases	Activity	Hypothesis	Huang, 2000; Lévárdy and Browning, 2005; Khalaf and Yang, 2006	0.2**

(*Continued*)

Table 7.11 (Continued) Lean activities and strategies that are positively correlated with mobile robot performance

Observation/hypothesis	Design phase	Design strategy/activity	Observation/hypothesis	Reference in literature to hypothesis	Percentage of variation in mobile robot performance (r^2)
13-STRA Aiming at striking a balance between fast response on one hand and stability, accuracy and payload fulfilment of the mobile robot final concept on the other	Conceptual and preliminary design phases	Strategy	Hypothesis	Yavuz, 2007; Fanuc Robotics, 2013	0.1**
19-ACT Adopting early verification and validation of the design concept, for example, early testing in the design process, to avoid becoming trapped in incompetent design concepts	Conceptual and preliminary design phases	Activity	Observation	N/A	0.1**
36-ACT Given the intertwined and overlapping nature of the subsystems of the mobile robot, for example, the mechanical–electronics interconnection for robot speed, adopting modular deliverables and testing, that is, testing deliverables of each module, rather than subsystems deliverables and testing	Preliminary and detailed design phases	Activity	Hypothesis	Clark and Baldwin, 2000	0.1**

(Continued)

Table 7.11 (Continued) Lean activities and strategies that are positively correlated with mobile robot performance

Observation/hypothesis	Design phase	Design strategy/ activity	Observation/ hypothesis	Reference in literature to hypothesis	Percentage of variation in mobile robot performance (r^2)
2-STRA Adopting the simplest design, which meets the design problem requirements specification using the minimum set and most effective combination of system components, rather than the cheapest and the lightest weight design	Conceptual and preliminary design phases	Strategy	Observation	N/A	0.1**
25-ACT Having a multidisciplinary team of novel designers, each of whom is aware of more than one of the relevant disciplines such as material science, design development approaches, leadership skills, and so forth, and is aware of the overlap and intersections between them, to improve the opportunity of achieving a better design and to minimise the risk of mistakes	Phase of design scope	Activity	Hypothesis	Amon et al., 1995	0.1**

(Continued)

Table 7.11 (Continued) Lean activities and strategies that are positively correlated with mobile robot performance

Observation/hypothesis	Design phase	Design strategy/activity	Observation/hypothesis	Reference in literature to hypothesis	Percentage of variation in mobile robot performance (r^2)
27-STRA Setting an operational design strategy of modular testing, that is, testing the deliverables between the system submodules as a way of verifying conformance of these system submodules to the conceptual functional requirements, to detect mistakes as early as possible and to minimise their impact on the successful completion of the design project	Conceptual and preliminary design phases	Strategy	Hypothesis	Lévárdy and Browning, 2005	0.1**
41-ACT Documenting the outcome of design discussions and consequently analysing that outcome	Conceptual and preliminary design phases	Activity	Hypothesis	Pahl and Beitz, 1998	0.1**
44-ACT Meeting collectively in the conceptual design phase rather than in detailed design phase	Conceptual and detailed design phases	Activity	Hypothesis	Court, 1998	0.1**
39-ACT Starting the design process as early as possible in the project time frame and assigning sufficient time for presenting the outcome of the design process milestones	Phase of design scope and conceptual design phase	Activity	Observation	N/A	0.1**

(Continued)

Table 7.11 (Continued) Lean activities and strategies that are positively correlated with mobile robot performance

Observation/hypothesis	Design phase	Design strategy/ activity	Observation/ hypothesis	Reference in literature to hypothesis	Percentage of variation in mobile robot performance (r^2)
54-ACT Checking accuracy of manufacturing and assembly of the final prototype, to avoid unexpected failure due to manufacturing defects and/or assembly mistakes	Detailed design phase	Activity	Hypothesis	Clark and Baldwin, 2000	0.1**
60-ACT Adopting quick testing of interdeliverables between the modules of the software draft code, and extensive testing of the overall software draft code on a prototype PCB or on an equivalent facility at the end of the project, to strike a balance between minimising the cost of the test and detecting mistakes as early as possible	Preliminary and detailed design phases	Activity	Hypothesis	Lévárdy and Browning, 2005	0.1**

** Correlation is significant at the 0.01 level (2-tailed).

The study has shown that 65% of typical mobile robot design activities and strategies are affiliated with the lean design paradigm, while the remaining 35% are affiliated with the agile design paradigm. In addition, it has been found with 99% confidence that 22% of the lean mobile robot design activities and strategies and 25% of the agile mobile robot design activities and strategies are among design activities and strategies that are significantly positively correlated with improved mobile robot performance. They have proved to significantly improve mobile robot performance by more than 10% and thus should receive the highest priority in assignment of design process resources, such as the resources indicated in Section 4.3. Owing to the usually limited resources available for a design project, there is a sort of trade-off and priority should be given to the design strategies/activities of higher correlation. The priority in resource allocation should be given to the design strategies/activities that have correlation r^2 higher than 0.1. A key reason why most of the design strategies/activities are weakly correlated with mobile robot performance is that they are necessarily basic strategies/activities that are commonly shared among designers. The design strategies/activities that have a higher correlation are among the differentiators between the highly successful mobile robots and the rest of the mobile robots. As can be gathered from Tables 7.11 and 7.12, the higher correlation was for the conceptual, preliminary and detailed design phases rather than the phase of design scope substantially because the conceptual, preliminary and detailed design phases cover most of the time span of the design project and a larger number of strategies so that they have higher likelihood of being more influential. A few design strategies/activities were expected to have a higher correlation, such as the sixth lean design strategy indicated in Table 7.11. Thus, the study has shown with 99% confidence that more than 10% of the variation in mobile robot performance can be explained by adopting a hybrid lean–agile mobile robot design paradigm that uses both lean and agile mobile robot design activities and strategies together in the mobile robot design process. Hence, adopting a hybrid lean–agile mobile robot design paradigm is technically valid.

7.10   Conclusions of the quasi-experiments

This book helps determine how and when value is added in designing integrated systems, by cost-effectively striking a balance between product development time and product performance attributes. To this end, this book proposes a design guideline to guide the product development process effectively by identifying the root causes of product success in the product development activities and strategies. To implement the design process effectively, this work identified how and when value is added in product design and development by determining the design activities and

Table 7.12 Agile activities and strategies that are positively correlated with mobile robot performance

Observation/hypothesis	Design phase	Design strategy/ activity	Observation/ hypothesis	Reference in literature to hypothesis	Percentage of variation in mobile robot performance (r^2)
1-STRA Shifting complexity towards the software subsystem, rather than the mechanical subsystem, for having the largest number of design iterations, if any, to occur within the software subsystem	Preliminary and detailed design phases	Strategy	Observation	N/A	0.3**
18-ACT Having iterations in the software, rather than in the mechanical subsystem, to achieve a shorter development time	Preliminary and detailed design phases	Activity	Observation	N/A	0.3**
1-STRA Having designs that are less vulnerable to failure modes (e.g. out-of-plane buckling, roll mode and pitch mode instability, and friction wear) (and consequently have better reliability) and are less exposed and less sensitive to the uncontrollable external factors (and consequently had better robustness), by shifting complexity to the software subsystem rather than to the mechanical subsystem	Conceptual and preliminary design phases	Strategy	Observation	N/A	0.2**

(Continued)

Table 7.12 (Continued) Agile activities and strategies that are positively correlated with mobile robot performance

Observation/hypothesis	Design phase	Design strategy/ activity	Observation/ hypothesis	Reference in literature to hypothesis	Percentage of variation in mobile robot performance (r^2)
1-STRA Having fewer constraints on the design by shifting complexity to the software subsystem and consequently to the virtual space, for example, data processing time, rather than to the mechanical subsystem and consequently to the physical space, for example, suspension space	Conceptual and preliminary design phases	Strategy	Observation	N/A	0.2**
12-ACT Adopting top-down system structure decomposition to analyse the functional structure of the mobile robot and consequently to map it to the requirements specification and consequently to come up with the mobile robot functional requirements	Phase of design scope	Activity	Hypothesis	Bernard, 1999	0.1**
26-STRA Having resource provisions for unforeseen problems, to minimise vulnerability of our design development process to the influence of external factors	Phase of design scope	Strategy	Hypothesis	Browning et al., 2005	0.1**

(Continued)

Table 7.12 (Continued) Agile activities and strategies that are positively correlated with mobile robot performance

Observation/hypothesis	Design phase	Design strategy/ activity	Observation/ hypothesis	Reference in literature to hypothesis	Percentage of variation in mobile robot performance (r^2)
31-STRA Adopting a decentralised decision making strategy, by empowering and authorising subteams to make tactical decisions without the need to refer them to the team leader, rather than strategic decisions that can degrade performance in this case	Conceptual, preliminary and detailed design phases	Strategy	Hypothesis	Krishnan, 1998; Pearce, 1999	0.1**
49-ACT Making a prediction of the progressive failure of the design through drawing sketches or conducting a finite element analysis, to avoid unexpected failure	Conceptual and Preliminary design phases	Activity	Observation	N/A	0.1**

** Correlation is significant at the 0.01 level (2-tailed).

strategies that are statistically the most and least correlated to product performance attributes and consequently product success.

The main objective of this book was to identify the engineering design activities and strategies that are critical to product success and differentiate successful from unsuccessful products, that is, the product success differentiators among design activities and strategies. This research goal was achieved by this work and the research criterion was met by finding answers to the secondary and consequently the primary research questions.

This book answered the three secondary research questions of this work. The research literature review answered the first secondary research question and identified the product attributes that should be investigated. The research literature review, hypotheses and experimental observations answered the second secondary research question and identified the design activities and strategies that impact upon them. Lastly, the research methodology and statistical analysis answered the third secondary research question and identified relationships between the product attributes and the design activities and strategies. Therefore, the primary research question in this work was answered and correlational and causal relationships between the technical attributes of product success and the product design and development activities and strategies were identified.

The proposed research methodology in this book was based on a research experimental approach and adopted hybridisation of both product-oriented and process-oriented design research perspectives to improve product performance attributes cost-effectively and consequently to make the product more successful in the market, and eventually to improve the industrial companies' profitability. The questionnaires, which were designed to collect data for this work, were designed with an emphasis on maximising the clarity of their wording and minimising the influence of the questionnaire's problems such as bias. To maximise clarity of questions, clarification footnotes were used. To minimise bias, a twofold strategy was adopted. First, to avoid researcher bias, closed questions were used; second, to spot respondent bias, inverted questions were repeatedly used. The statistical sampling in this work was representative in terms of sampling design that was suitable for generalisation with cost-effectively fair statistical results and sample size that satisfied the minimum statistically representative sample size. In addition, the data used in this study satisfied fairly the three aspects of a statistical power test. First, the statistical significance aspect was satisfied in terms of proved significance of research results at a 2-tailed 0.01 level; second, the size of the population was reasonably satisfied because despite a large population size that can negatively affect the data statistical power, probability sampling design was used to mitigate such a negative effect; third, the reliability of the data was fairly satisfied through passing the data reliability and verification test in the Section 6.7 in Chapter 6. In addition, the research results were validated successfully in terms of the four validity types: first, in terms of statistical conclusion validity, as the resulted relationships were meaningful and reasonable; second, in terms of internal validity, as the results were causal rather than just descriptive; third, in terms of construct validity, as the results represented what is theoretically intended; fourth, in terms of external validity, as the results could be generalised to the population. A case study conducted in this context showed that the companies that adopt the hybrid approach become much more successful than those that do not.

By reviewing the relevant literature, it was found that there are six approaches to understanding and managing product design and development process: (1) strategic and marketing-oriented perspective; (2) cost, accounting and financial-oriented perspective; (3) product development time-oriented perspective; (4) social sciences perspective; (5) technical-oriented perspective and (6) hybrid approach of both strategic and technical approaches. It was also found that the hybrid approach is the most comprehensive one to strike a balance integratively among the five pillars of product development: product attributes, product development time, product development cost, customer value and enterprise business goals. It was found that an integrated design system has five subsystems: system design, project management, mechanical design, electronic design and software design. Across these five subsystems, the positively and the negatively correlated sets of design activities and strategies with product success were empirically and statistically identified.

Product design and development project resources and provisions should be allocated based on the respective most important design activities and strategies. There were 21 design activities and strategies that had the most significant positive influence, more than 10% (at the 2-tailed 0.01 level), on the variations in the technical attributes of product success; hence, they should receive the most focus and highest priority in resource allocation because they are product success differentiators. The first 8 of these 21 design activities and strategies were affiliated with the system design dimension and were the following:

1. In the integrated design system, shifting complexity towards the software, rather than the mechanical subsystem, to have the largest number of design iterations, if any, to occur within the software subsystem, followed accordingly by the electronics subsystem number of iterations
2. In the integrated design system, having iterations in the software, rather than in the mechanical subsystem, to achieve a shorter development time

3. In the integrated design system, considering reliability of the final product in the design process, in terms of the ability of the product to perform its required functions under stated conditions for a specified product service time

4. In the integrated design system, having designs that are less vulnerable to failure modes (e.g. out-of-plane buckling, roll mode and pitch mode instability and friction wear) (and consequently have better reliability) and are less exposed and less sensitive to the uncontrollable external factors (and consequently had better robustness), by shifting complexity to the software rather than to the mechanical subsystem

5. In the integrated design system, having fewer constraints on the design by shifting complexity to the software subsystem and consequently to the virtual space, for example, data processing time, rather than to the mechanical subsystem and consequently to the physical space, for example, suspension space

6. In the integrated design system, aiming at striking a balance between fast response on one hand and stability, accuracy and payload fulfilment of the product final concept on the other

7. In the integrated design system, adopting top-down system structure decomposition to analyse the functional structure of the product and consequently to map it to the requirements specification and consequently to come up with the functional requirements of the product

8. Adopting the simplest design, which addresses the design problem requirements specification by the minimum set and most effective combination of system components, rather than the cheapest and the lightest weight design

The next 10 of these 21 design activities and strategies were affiliated with the project management dimension and were the following:

1. In the integrated design system, adopting testable design interdeliverables within and among system modules based on project milestones, to detect mistakes as early as possible and to minimise their impact on the successful completion of the design project

2. In the integrated design system, adopting early verification and validation of the design concept, for example, early testing in the design process, to avoid becoming trapped in incompetent design concepts

3. In the integrated design system, given the intertwined and overlapping nature of the subsystems of the product, for example, the mechanical–electronics interconnection for robot speed, adopting modular deliverables and testing, that is, testing deliverables of each module, rather than subsystems deliverables and testing

4. Having resources provisions for unforeseen problems, to minimise the vulnerability of our design development process to the influence of external factors

5. In the integrated design system, having a multidisciplinary team of designers, each of whom is aware of more than one of the relevant disciplines such as material science, design development approaches, leadership skills and so forth, and is aware of the overlap and intersections between them, to improve the opportunity of achieving a better design, and to minimise the risk of mistakes

6. In the integrated design system, setting an operational design strategy of modular testing, that is, testing the deliverables between the system submodules as a way of verifying conformance of these system submodules to the conceptual functional requirements, to detect mistakes as early as possible and to minimise their impact on the successful completion of the design project

7. In the integrated design system, adopting a decentralised decision-making strategy, by empowering and authorising subteams to make tactical decisions without the need to refer them to the team leader, rather than strategic decisions that can degrade performance in this case

8. In the integrated design system, documenting the outcome of design discussions and consequently analysing that outcome

9. In the integrated design system, meeting collectively in the conceptual design phase rather than in the detailed design phase

10. In the integrated design system, starting the design process as early as possible in the project time frame and assigning sufficient time for presenting the outcome of the design process milestones

The next 2 of these 21 design activities and strategies were affiliated with the mechanical design subsystem dimension and were the following:

1. Checking the accuracy of manufacturing and assembly of the final prototype, to avoid unexpected failure due to manufacturing defects and/or assembly mistakes

2. Making a prediction of the progressive failure of the design through drawing sketches or conducting a finite element analysis, to avoid unexpected failure

The last of these 21 design activities and strategies was affiliated with the software design subsystem dimension and was the following:

1. In the integrated design system, adopting quick testing of interdeliverables between the modules of the software draft code, and extensive testing of the overall software draft code on a prototype PCB

or on an equivalent facility, to strike a balance between minimising cost of test and detecting mistakes as early as possible

On the other hand, this work pointed out that the design activity and strategy that has the most significant negative influence on the variations in the technical attributes of product success at the 2-tailed 0.01 level was the following:

1. In the integrated design system, testing aggregately the software sub-system at the end of the project rather than adopting quick testing

Hence, it should receive the lowest priority in assignment of resources.

It was found also that there are 24 uncorrelated design activities and strategies with technical attributes of product success that are adopted by all designers as they are important to success but they are not success differentiators (at the 2-tailed 0.01 level). These should receive a low focus and low priority in resource allocation. The first 5 of these 24 design activities and strategies were affiliated with the system design dimension and were the following:

1. In the integrated design system, considering safety of use of the final product in the design process, in terms of accident prevention, risk identification and risk control and/or minimisation
2. In the integrated design system, adopting a systematic design approach, that is, (1) clarifying the design problem; (2) identifying requirements specification and design constraints; (3) sorting the requirements specification in a ranked order; (4) identifying functional requirements, for example, using an axiomatic design methodology and identifying core design values; (5) constructing a table of design options based on brainstorming; (6) making concept evaluation, for example, based on a weighted criteria evaluation matrix and (7) making an overall strategic decision of the conceptual design
3. In the integrated design system and in case of adopting the systematic design approach, paying special attention to the quality of performing each of the steps of the systematic design approach
4. In the integrated design system, setting an error-proof operational design strategy, for example, constructing 3D CAD model of the system assembly drawing to minimise waste of time and resources that is due to making avoidable mistakes
5. In the integrated design system, aiming at striking a balance between maximising functionality of the product final concept on one hand and minimising waste of time and resources by minimising design iterations on the other

The next 12 of these 24 design activities and strategies were affiliated with the project management dimension and were the following:

1. Adopting cross fertilisation of design concepts among the design team, such as sharing ideas and/or modifying each other's ideas
2. In the integrated design system, adopting live meetings of the design team in the conceptual and preliminary design phases, and e-mails in the detailed design phase, as the principal communication method, to strike a balance between the synergy of face-to-face meetings on one hand and the time effectiveness and message delivery effectiveness of the electronic means of communication, such as e-mails, on the other
3. Adopting both sequential and concurrent design activities according to the nature of each task and the preceding relationships between them, to maximise utilisation of resources and shorten the design development time
4. In the integrated design system, being more interested in meeting project delivery deadlines than in achieving the project objectives to a specified cost
5. In the integrated design system, adopting redeployment of the human resources of the design team to cope with fluctuations in the workload of project activities
6. In the integrated design system, conducting a design research activity (e.g. research literature review, review of previous student reports, etc.) and/or Internet literature review search to become acquainted with what other designers have done to design and implement similar products and to become acquainted with state of the art of the design of the required product
7. In the integrated design system, considering the project total time duration as the most important constraint and consequently being more interested in the critical path, the longest sequence of time-constrained design tasks in the project plan, than in the critical chain, the longest sequence of resource-constrained design tasks in the project plan
8. In the integrated design system, having experienced designers in the design team, to better conceive the assumptions and constraints accompanied by similar designs
9. In the integrated design system, having experienced designers in the design team, to have a foolproof design process and to prevent avoidable mistakes in the design process, such as mounting the more heat-sensitive electronic components first on the PCB prototype
10. In the integrated design system, selecting a team leader with a software specialisation and familiarity with electronics and mechanical subsystems

11. In the integrated design system, estimating the duration of each project activity based on work breakdown structure, experience and comparison with comparable projects
12. In the integrated design system, coordinating between the design sub-teams concerning the interrelated elements of the system, for example, software–electronics interconnection for microchip pin allocation and for positioning of sensors that fits all the required routing

The next 3 of these 24 design activities and strategies were affiliated with the mechanical design subsystem dimension and were the following:

1. Conducting a to-scale cardboard prototype, mock-up or equivalent in the mechanical subsystem
2. Building a to-scale design prototype to be tested in the real operational environment in the detailed design phase
3. In the integrated design system, determination by the mechanical subteam of the optimal number of degrees of freedom of the system, for example, of picking-up and unloading mechanisms, that fulfil the functional requirements and minimise the number of needed controllable mechanisms

The next of these 24 design activities and strategies was affiliated with the electronics design subsystem dimension and was the following:

1. In the integrated design system, developing the electronics circuitry such that it minimises the number of required reads and writes, and maximises utilisation of channelling and port addressing

The last 3 of these 24 design activities and strategies were affiliated with the software design subsystem dimension and were the following:

1. In the integrated design system, aiming at implementing explanatory annotation of the software code for purposes of software code debugging and maintenance
2. In the integrated design system, adopting a parameterisation design strategy, that is, changing the value of one parameter changes the values of many related parameters accordingly, to facilitate software code debugging, maintainability and/or scalability of the final concept of the product
3. In the integrated design system, spending at least 10% of the total time duration of the software subsystem on functional analysis and software architecture, for example, making program flow chart, rather than on the programming code implementation itself, to minimise waste of time and resources and to eliminate sources of avoidable mistakes

7.11   Summary and major contributions

This book addressed current industrial needs by helping to determine how and when value is added in designing integrated systems. It aimed at improving industrial companies' profitability through cost-effective management of the product development design activities and strategies. It is affiliated with the engineering design management research area. It attempted to apply Pareto's phenomenon, that is, the 80/20 rule, to the product design and development activities and strategies by identifying the vital few product design and development activities and strategies as the success factors that account for a high percentage of the product success. The book proposed a design guideline for the product design and development process through exploring the correlation and causality between the design process activities and strategies on the one hand and the product attributes and product cost-effectiveness and consequently a company's profitability on the other.

The proposed research methodology in this book was based on a research experimental approach. This book adopted hybridisation of both product-oriented and process-oriented design research perspectives to improve product performance attributes cost-effectively and consequently to make the product more successful in the market, and eventually to improve the industrial companies' profitability. In addition, the hybrid approach to understanding and managing the product development process was employed in this work, as it represents the most comprehensive approach in this context. In addition, the research methodology of this research book adopts hybridisation of all three major research methods: exploratory (causal), descriptive, and correlational research because the research needed some tools of each of these. For instance, from the exploratory research, the quasi-experiment was used; from the descriptive research, the frequency analysis was needed and from the correlational research, the causal comparative observational research and the archival research in terms of literature review were employed. Moreover, this book adopted the hybrid quantitative–qualitative research approach because the research needed some tools of both of these; for instance, from the quantitative research, questionnaires and statistics were employed; from the qualitative research, observational research was used. The statistical analysis approach was implemented in this book to rigorously analyse the research data and verify and validate the research results. To validate the research observations further, they were included in the research questionnaires.

The book could statistically identify the design activities and strategies most and least correlated to product performance attributes and consequently product success. This book answered the three secondary research questions of this work as follows:

1. The research literature review answered the first secondary research question and identified the product attributes that should be investigated.
2. The research literature review, hypotheses and experimental observations answered the second secondary research question and identified the design activities and strategies that impact upon them.
3. The research methodology and statistical analysis answered the third secondary research question and identified relationships between the product attributes and the design activities and strategies. Therefore, the primary research question of this book was answered by identifying correlational and causal relationships between the technical attributes of product success and the product design and development activities and strategies.

The major contributions of this book are as follows:

1. Empirical investigation of the internal aspects of the design process of mobile robots through identifying the most statistically influential mobile robot design strategies and activities on mobile robot performance
2. Identification of the dimensions and subsystems of the mobile robot design process
3. Identification of the level of product performance that yields maximum return on complex product development
4. Identification of the key approaches to understanding and managing the complex product design process
5. Proposing an approach to effectively manage risk in the complex product design process that hybridises attributes of both the lean and agile design paradigms
6. Identification of the key determinants of the mobile robots development process

chapter eight

Concluding remarks and future research directions

8.1 Concluding remarks

The book described an empirical investigation-based design guideline that strikes a balance between mobile robot development time and performance, optimising the added value in the mobile robot design process. The book presented the most efficient among the key approaches to understanding and managing the product design process. An approach to manage risk in the product design process effectively was also proposed. This design guideline can help novel designers in cost-effectively managing the mobile robot design process, enhancing their competitiveness. It is limited to complex products in the structural design and construction engineering and robotics sectors.

8.2 Directions for future research

The field of product design is growing and moving towards a well-established discipline. The growth of any discipline relies mainly on the accumulation of research-based knowledge. The product design discipline is no exception. At the heart of this progress lies the complex product development process. Lean design remains a key research direction in engineering design. Designing complex products based on lean measures is a key approach in this direction (Emmitt, 2011; Tribelsky and Sacks, 2011). In addition to the lean measures, flexible solutions emerge as a key approach in this direction as well (Hansen and Olsson, 2011). A promising approach that emerged partly from the lean direction is the design concept of first and last value, which helps in maximising value delivered to customers through design.

Investigating whether or not and to what extent pure product innovators have an advantage over pure process innovators has been drawing interest. Applying the quality function deployment (QFD) method during the early phases of a product development process to facilitate collaborative design concept evaluation has been increasingly gaining interest. Utilising the QFD method throughout an iterative design process without introducing too much complexity to the agile development process is thus a promising research direction (Benker and Raduma, 2014). Operational agility is a capability that enables design firms to sense changes in turbulent business environments and to take appropriate actions to seize market opportunities (Huang et al., 2014). Measurement of information sensitivity, synergy and fluidity in the process of developing information processing capability for operational agility is a promising research area. Order fulfillment is a key process in managing the supply chain of complex products. Examining how the design firm develops information processing capabilities to achieve operational agility throughout the order fulfillment process is a related promising area.

Integrating design at a strategic level within the company, shifting the perspective of design from a product focus towards a strategic focus, emerges as another promising research direction. A key approach in this direction includes how a company transitions from an exclusively product-focused utilisation of design to a process-level application of design (Doherty et al., 2014). Setting a strategic design intent to help designers become flexible around different perspectives and approaches such as core design approaches, for example, iteration, visualisation, prototyping and sharing is a promising sub-direction (Wildman, 2014). Transformation of the traditional design business model, that is, problem identification and solution, to one in which design thinking is employed through one of these core design approaches is another promising sub-direction (Badding et al., 2014). The intersection of processes of complex products creation and brand management activities is another promising approach in this direction (Santos and Morillo, 2014). For fostering innovation, one of the obstacles is the uncertainty over the process, and the core design approach of prototyping has been acknowledged as a key element of the design methodology to embrace the uncertainty. Developing a theoretical model of the business model prototyping with the four key elements of iterative and agile learning, tangibility, complexity and synthesis is also a promising research sub-direction (Amano, 2014).

Design for sustainability is another research direction that is increasingly gaining interest. Sustainably devising complex products that respond to environmental concerns is a cornerstone in this research direction. User-centered design practices including timelessness, innovation, durability and craftsmanship should meet the goals of sustainability, which include meeting environmental legislation requirements, achieving profitable growth and meeting customer expectations. Promising approaches in this direction include

- Innovating low-carbon sustainable solutions (Simeone, 2014)
- Investigating the development of eco-innovation, green product innovation and environmental innovation in small and medium-sized enterprises (SMEs)

- Exploring the link between leadership and types of behaviours of the design team and research and development collaboration for environmental innovation
- Investigating the horizontal development of leadership, for example, new skills, abilities and behaviours, and vertical development of leadership, for example, expanding the mindset capacity and improving the mindset structure, of the members of the design team at the design firms that are specialised in complex products
- Investigating the relationship between the key design thought methods, need-finding, brainstorming and prototyping, and the level of design experience of the team members who design complex products (Seidel and Fixson, 2013)
- Exploring the link between cross-functional collaboration and the financial success of new complex products (Graner and Mißler-behr, 2014)
- Investigating the relationship among the collaboration of innovation intermediaries, vertical integration of design firms and market success of complex products
- Exploring the relationship among complex products development portfolio, product launch rate and design firm profitability (Green and Raman, 2014)
- Investigating internal networks among the design teams of innovation firms
- Investigating the success factors on innovativeness in complex products development

Understanding better the relationship between product architectures and organisation structures is a promising area of research. A design structure matrix (DSM) can help in comparing and contrasting alternative product and organisational configurations (Browning, 2001; Sosa et al., 2004). The link between process and organisation structures, and the corresponding relationship between the activity- and team-based DSMs, is also a promising area. Investigating the propagation of error in an assembly using a DSM for process improvement is a promising area as well (Browning, 2001). Investigating the use of Cambridge Advanced Modeller (CAM) to determine which tasks within each phase of a complex project should or should not be performed concurrently has been gaining research interest (Maier et al., 2014). Such a modeller can help in identifying measures or constructs to quantify the dependencies between the design tasks or subsystems. This can lead to accelerating complex products development, improving complex products quality, reducing complexity and managing design iteration strategically. Investigating how to build a process model incrementally for incomplete specifications during the early stages of the modelling activities when it is impossible or inconvenient to make the specifications precise and exhaustive is a related promising research direction.

Another research direction is creation of ecosystems of interconnected products, services and solutions that can be accessed wherever and whenever users desire. Design thinking by driving innovation through true collaboration from the onset is a promising way to create such ecosystems (Gardien et al., 2014). Investigating why the companies that adopt inside-out open innovation are more likely to create radical innovations and tend to sell a greater number of new products than the companies that adopt closed innovation is a promising research area (Inauen and Schenker-Wicki, 2012). An in-depth study of challenges related to the transition from one type of customer involvement in open innovation in turbulent environments to another is another promising area.

Knowledge management remains a key research direction in the design management of complex products. This direction also supports sustainability (Abbas et al., 2011). Dynamic knowledge management through combining dynamic capabilities and resources along with internationalisation and networking processes to make the design firms more competitive is a promising approach in this direction. Design organisations strategically compete on four generic capabilities: cost efficiency, quality of products, speed of delivery and flexibility of operations. Investigating whether to focus on one or on multiple capabilities simultaneously is a promising research sub-direction (Ashwini Nand et al., 2014). In investigating the methodology of evaluating innovation capabilities of design firms that specialise in designing complex products, using a fuzzy system is another promising research sub-direction.

Complex products are becoming more multidisciplinary. Their boundaries have been expanding from a product as a package to a product-service system with life cycle aspects through the knowledge-centered integration of product development. Addressing these life cycle aspects in the development of complex products leads to product development with better quality, lower costs, quicker delivery and greener performance (Tomiyama and Meijer, 2006). To achieve these goals, investigating the integration of different types of knowledge about the life cycle of complex products is a promising research area. This integration of knowledge may include concurrently integrating different activities taking their fields, for example, mechanical design, electronics design and software design, into consideration during development of complex products. The seamless integration of these design activities can result from sharing a common database and proper feedback and/or feedforward mechanisms. Investigating the resource-based view (RBV) of the design firm focussing on the relationship between innovation and firm performance on the one hand and resources and firm performance on the other simultaneously is a promising research direction (De Zubielqui et al., 2014). Design firms usually experience environmental turbulence in which remaining strategically flexible is of crucial importance to ensure survival. Developing dynamic capabilities of design firms to meet sustainable development challenges is thus also a promising research direction (Schneider and Spieth, 2014). Exploring how market orientation and technology orientation within the fuzzy front end phase affect success of product innovation in complex products is a promising research direction as well (Liu and Su, 2014). Investigating the relationship among reverse innovation, new complex product performance and type of markets (developed or emerging market) has been receiving research interest. In an endeavor to develop process capability, it would be interesting to investigate using DSM how mature processes deal with novel complex product development. Comparing a component-based to an activity-based DSM can help in exploring the benefits of architectural modularisation to the design process of complex products. Investigating system dynamic modelling of the complex products development process is a promising area. This dynamic modelling manipulates the complex products development process as a set of hierarchies of subsystems interconnected at different levels: industry, product, organisation and process. Therefore, the interactions between resources allocation and design tasks rates and the necessary trade-offs to achieve the targets in terms of deadline, budget and quality can be captured.

Among the research directions in the complex product development process is integrated design. Information modelling is one of the pillars of this direction, ensuring smart integration of the elements of the design process for achieving the design targets (Owen et al., 2011). Investigating technology road-mapping and its strategic value in exchanging information among the members of the design team is a promising research subdirection, as are the interactional aspects of communicating innovative ideas. The value of the co-creation process for the development of new complex products is a promising approach. This value includes leveraging innovative design capability, disseminating knowledge outcome and boosting knowledge management capability and providing adaptive governance capability. The level of innovativeness of the design project is an important determinant of product potential that demands functional and integrative communication capabilities of design firms. Investigating these capabilities is a promising approach in this direction (Tepic et al., 2013). Investigating the cyclic innovation model (CIM) as an effective framework for understanding and managing the innovation process on complex products is a promising research sub-direction. The integration of customers in the co-design process is one of the most important aspects of mass customisation. Investigating their role as co-designers and the entailed integration capability and how they influence the configuration of their products is a promising research area (Theilmann and Hukauf, 2014).

Distributed development of complex products often involves hierarchically organised work. Investigating the relationship between information visibility and the software development pattern in the distributed development of complex products is a promising research area. A related research area is Business Infrastructure Management and/or Business Information Management (BIM). A promising research direction in this area is Business IT Management (Elmualim and Gilder, 2013), which improves the effectiveness of the design management of complex products. A collective decision-making support system for innovation management such as online systems created to facilitate the participation of professionals in the decision-making processes of innovation development is a promising approach. Investigating the role of virtual networks and cloud computing strategy in complex products design management is a promising related area. Design practices have evolved to include shaping the long-term identity of organisations. Hence, a key research focus is on innovations that produce not just new products, but also strategically build resilient and adaptive organisations that over time can leverage change to renew competitive advantages. Considering design as a strategic investment is a promising approach in this research area, for instance exploring (1) the role of design and design management in supporting the corporate strategy, (2) the link between design management and product sustainability and (3) how design can be quantified when measuring brand equity and loyalty (Jenkins and Golsby-Smith, 2013). Building brand equity for performance through design is hence a promising approach in this direction. Entrepreneurial design is also a promising approach. Strategically investigating the development of incremental innovation versus radical/disruptive innovation taking into consideration that corporate core-capabilities can hinder an organisation's adaption to innovation has been receiving interest (Hamada, 2014). The acceleration of innovation in the old technology in response to the threat from the new technology to revitalise old technology-based complex products has been gaining interest as well.

Another research direction is quantitatively investigating the link between design management and process. Key aspects of this direction include

1. What are the optimal processes and methodologies for objective setting and metric definition?
2. What key performance indicators should be measured and managed, and which tools should be used (Jenkins and Golsby-Smith, 2013)?

Integrating products and services into a single solution is also a promising research direction. Smart product-service systems (Smart PSSs) allow organisations to develop relations with customers in new ways and have a growing presence in the marketplace. An understanding of the challenges emerging from the integration of product and service is of increasing relevance for the effective management of the design process (Valencia et al., 2014).

Designers usually do not change simply because design processes change. Managing organisational culture change towards hybrid lean–agile design of complex products in design firms thus can be the key challenge in the implementation of such a design paradigm. Investigating empirically such implementation in industry is a promising research direction. For instance, improving the testing tools and procedures used for software testing such as shifting towards automated testing of developed software when transitioning to the hybrid lean–agile design paradigm in the design process of mobile robots is a key part of such a challenge. Disciplined experimentation to identify quickly the key needs of the customers, make fast prototyping and simulate a realistic customer journey directly with the end customers has been proposed to reduce innovation-related risk (Sola et al., 2014). Extending the value of the design by incorporating uncertainty into the design management process is a promising approach in this direction, as is investigating the relationship between uncertainty and radical innovation. Perceiving and managing risk in the development process of complex products may gain future interest, particularly with respect to the risk-taking design strategy. Whether and how risks in the development process of complex products are interlinked is a promising related research area (Jerrard et al., 2008). Investigating the methods of identifying technical and market risks throughout the development process of complex products rather than only in the earlier phase is a promising area.

Design is strategically a major contributor to the formation of brand identity. Identity formation in relation to design and brand strategy is hence a promising research direction. Design for building a strong brand and design for challenging status quo are among promising approaches in this direction. A related direction is how profitable solutions and value can be created from intangible aspects and customers' emotions. Investigating how intangible resources can be drivers of high growth of design firms that specialise in complex products is a related promising research direction as well. Developing design tools to this end to facilitate the process of innovation in design firms is a promising approach, as is designing with customers through Web-based methods (Straker and Wrigley, 2014). A user-centered approach in the early stages of the design process is increasingly gaining interest.

Significant disruption to existing design business models is driven primarily by (1) the pace of technological change and innovation and (2) changing expectations on the part of customers. Developing strategic business models for innovation based on applying design-concept methodologies and approaches to these strategic problem spaces is receiving increasing interest (Jenkins and Fife, 2014). The investigation of the relationship between existing complex products and adoptions of substituting technologies has been gaining research interest as well. Investigating the influence of knowledge-based change on technological capability and performance of complex products design firms in times of increasing uncertainty is a promising research direction.

Appendix A: Design management research questionnaire on the integrated design project

The aim of this questionnaire is to explore, for research purposes, the hypothesised correlation and causality between design process activities and strategies on the one hand and the quality of the final product performance on the other. To explore this hypothesised correlation, we would like to collect some data from designers about their real design process activities and strategies. Consequently, we would like you to kindly answer this questionnaire. Thank you in advance for your cooperation.

Please, tick (✓) the **most appropriate evaluation category** to what your design team actually adopted and conducted in the design process.

Design team number

Design process activities/strategies	Strongly agree	Agree	Do not know	Disagree	Strongly disagree
System design dimension					
1. We aimed at shifting the design complexity towards the software subsystem, rather than the mechanical subsystem.					
2. We aimed at the simplest design rather than the cheapest design and the lightest design.					
3. We chose the mechanical design concept first, then the electronics concept, and then the software concept.					
4. We adopted electronic distinguishing, that is, sensing, components rather than mechanical counterparts.					
5. We adopted only standard components.					
6. We adopted a modular design that is, splitting the overall system into modules that interface with each other, to speed up the design development process, facilitate error-tracing and maintainability and minimise the impact of design iterations.					
7. We adopted a systematic design approach, that is, (1) clarifying the design problem; (2) identifying requirements specification and design constraints; (3) sorting the requirements specification in a ranked order; (4) identifying functional requirements and core design values; (5) constructing a table of design options based on brainstorming; (6) making concept evaluation, for example, based on a weighted criteria evaluation matrix and (7) making an overall strategic decision of the conceptual design.					

(*Continued*)

Design team number (Continued)

Design process activities/strategies	Strongly agree	Agree	Do not know	Disagree	Strongly disagree
8. In case of adopting the systematic design approach, we paid attention to the quality of performing each of the steps of the systematic design approach.					
9. We set an error-proof operational design management strategy for the design project.[a]					
10. We adopted a tolerant design that can accommodate variations in the inputs to the system.					
11. We aimed at striking a balance between maximising functionality of the product final concept and minimising design iterations.					
12. We adopted top-down system structure decomposition to analyse the functional structure of the product to map this functional structure of the product to the requirements specification and consequently to come up with the product functional requirements.					
13. We aimed at striking a balance between fast response on one hand and stability, accuracy and payload fulfilment of the product final concept on the other.					
14. We considered reliability of the final product in the design process, in terms of the ability of the product to perform its required functions under stated conditions for a specified product service time.					
15. We considered maintainability of the final product in the design process, in terms of the ease with which maintenance of the product functional units can be performed.					

(Continued)

Design team number (Continued)

Design process activities/strategies	Strongly agree	Agree	Do not know	Disagree	Strongly disagree
16. We considered safety of use of the final product in the design process, in terms of accident prevention, risk identification and risk control and/or minimisation.					
17. We predicted working scenarios of risk, for example, inappropriate lighting conditions, or missing a junction on the route, and prepared accordingly action plans, for example, error-recovery tactic of tolerant triggering threshold, or memorising last state of all input signals and correcting path accordingly based on the expected milestone on the route plan, to improve performance robustness of the final concept of the product.					
18. We had the largest number of design iterations, if any, to occur within the software subsystem.					
Project management dimension					
19. We conducted early verification and validation of the design concept, that is, early testing of the design concept.					
20. We selected a design team leader with software specialisation and familiarity with electronics and mechanical subsystems. Which design subteam was the team leader affiliated with? (Mechanical, Electronics, or Software)					
21. We spent more time on the conceptual and preliminary design phases together than on the detailed design phase.					

(Continued)

Design team number (Continued)

Design process activities/strategies	Strongly agree	Agree	Do not know	Disagree	Strongly disagree
22. We visited and reviewed similar designs that solved similar design problems.					
23. We adopted cross fertilisation of design concepts among the design team, such as sharing ideas and/or modifying each other's ideas.					
24. We assigned at least 5% of the total time duration of the project to make a macroscale aggregate plan in the scoping phase, and had resource provisions, for example, time buffer, for variations in details, and made detailed design plan at the preliminary design phase. What is the approximate percentage that you assigned?					
25. We had a multidisciplinary team of designers, each of whom is aware of more than one of relevant disciplines such as material science, design development approaches, leadership skills, and so forth, and is more aware of the overlap and intersections between them, to improve the opportunity of achieving a better design, and to minimise the risk of mistakes.					
26. We had resources provisions, such as buffer time, for unforeseen troubles, to minimise vulnerability of our design development process to the influence of external factors.					
27. We set an operational design strategy of modular testing, that is, testing the deliverables between the system submodules as a way of verifying their conformance to the conceptual functional requirements, to detect mistakes as early as possible and to minimise their impact on the successful completion of the design project.					

(Continued)

Design team number (Continued)

Design process activities/strategies	Strongly agree	Agree	Do not know	Disagree	Strongly disagree
28. We made more iterations in the conceptual and preliminary design phases than in the detailed design phase, to increase the likelihood of achieving a better design, while minimising waste of time and resources.					
29. We adopted testable design interdeliverables within and among system modules based on project milestones, to detect mistakes as early as possible and to minimise their mistakes on the successful completion of the design project.					
30. We adopted live meetings of the design team in the conceptual and preliminary design phases, and e-mails in the detailed design phase, as the principal communication method, to strike a balance between the synergy of face-to-face meetings on one hand and the time effectiveness and message delivery effectiveness of the electronic means of communication, such as e-mails, on the other.					
31. We adopted a decentralised decision-making strategy, by empowering and authorising subteams to make tactical decisions without need to refer them to the team leader.					
32. We adopted both sequential and concurrent design activities according to the nature of each task and the preceding relationships between them, to maximise utilisation of resources and shorten the design development time.[b]					

(Continued)

Design team number (Continued)

Design process activities/strategies	Strongly agree	Agree	Do not know	Disagree	Strongly disagree
33. Our design team was more interested in meeting project delivery deadlines than in achieving the project objectives to a specified cost.					
34. Our design team estimated the duration of each project activity based on work breakdown structure, experience and comparison with comparable projects.[c]					
35. Our design team adopted redeployment of the human resources of the design team to cope with fluctuations in workload of project activities.					
36. Given the intertwined and overlapping nature of the subsystems of the product, for example, the mechanical–electronics interconnection for robot speed, our design team adopted modular deliverables and testing, that is, testing deliverables of each module, rather than subsystems deliverables and testing.					
37. Our design subteams coordinated with each other concerning the interrelated elements of the system, for example, software–electronics interconnection for microchip pin allocation and for positioning of sensors that fits all the required routing.					
38. Our design team conducted a design research activity (e.g. research literature review, review of previous student reports, etc.) and/or Internet literature review search to become acquainted with what other designers did to design and implement similar products and to become acquainted with the state of the art of the design of the required product.					

(Continued)

Design team number (Continued)

Design process activities/strategies	Strongly agree	Agree	Do not know	Disagree	Strongly disagree
39. Project total time duration was the most important constraint, and accordingly we started our design process as early as possible in the project time frame and assigned sufficient time to presenting the outcome of the design process milestones, without interrupting the design process.					
40. We considered the total time duration of the project as the most important constraint and consequently were more interested in the critical path, the longest sequence of time-constrained design tasks in the project plan, than in the critical chain, the longest sequence of resource-constrained design tasks in the project plan.					
41. We did not document the outcome of our design discussions and consequently did not analyse that outcome.					
42. We believed that the greater the number of experienced designers in the design team, the better they conceive the assumptions and constraints accompanied by similar designs.					
43. We believed that the greater the number of experienced designers in the design team, the more they prevent avoidable mistakes in the design process, and the more their design process becomes foolproof.					
44. We believed that the more the design team met collectively in the conceptual design phase than in detailed design phase, the more the likelihood of achieving a more effective design concept.					

(Continued)

Design team number (Continued)

Design process activities/strategies	Strongly agree	Agree	Do not know	Disagree	Strongly disagree
45. Our design subteams met collectively in the conceptual design phase to choose the mechanical and electronics design concepts, respectively.					
46. Our electronics and software design subteams conducted initial test experiments in the preliminary design phase on the chosen conceptual mechanical and electronics design components to explore the full potential and scope of these components.					
Mechanical subsystem dimension					
47. We constructed a to-scale cardboard prototype, or mock-up, in the mechanical subsystem.					
48. We adopted strength-adaptable mechanical chassis design approach that can embrace further changes, for example, future changes in electronic components weight and layout, to improve agility of the design process.					
49. We made a prediction of the progressive failure of the design through drawing sketches or through conducting a finite element analysis, to avoid unexpected failure.					
50. We made sketches to generate concepts in the conceptual design phase.					
51. We constructed CAD models in the preliminary design phase.					

(Continued)

Design team number (Continued)

Design process activities/strategies	Strongly agree	Agree	Do not know	Disagree	Strongly disagree
52. We built a to-scale design prototype to be tested in the real operational environment in the detailed design phase.					
53. We considered manufacturability starting from the conceptual design phase onwards in the design process, to minimise waste of time and resources and to eliminate relevant avoidable mistakes and unnecessary design iterations.					
54. We checked accuracy of manufacturing and assembly of the final prototype, to avoid unexpected failure due to manufacturing defects and/or assembly mistakes.					
55. Our mechanical subteam determined the optimal number of degrees of freedom of the system, foe example, of picking-up and unloading mechanisms, that fulfil the functional requirements and minimise the number of needed controllable mechanisms.[d]					

Electronics subsystem dimension

56. We set an error-tracing operational design strategy, such as coloured electronic wiring in the electronics subsystem, to facilitate iterations, troubleshooting, software code debugging, and/or maintainability of the final concept of the product.					
57. Our electronics designers chose the most reliable types of sensors and extended the sensors' robustness through software manipulation.					

(Continued)

Design team number (Continued)

Design process activities/strategies	Strongly agree	Agree	Do not know	Disagree	Strongly disagree
58. Our electronics designers developed the electronics circuitry such that it minimises the number of required reads and writes and maximises utilisation of channelling and port addressing.					
Software subsystem dimension					
59. We spent at least 10% of the total time duration of the software subsystem on functional analysis and software architecture, for example, making a program flow chart, rather than on the programming code implementation itself, to minimise waste of time and resources and to eliminate sources of avoidable mistakes. What is the approximate percentage you spent?					
60. We adopted quick testing of interdeliverables between the modules of the software draft code and extensive testing of the overall software draft code on a prototype printed circuit board (PCB) or on an equivalent facility, to strike a balance between minimising cost of test and detecting mistakes as early as possible.					
61. We assigned at least 10% of the total time duration of the software subsystem to the overall testing time, that is, of both quick testing and extensive testing. What is the approximate percentage you assigned?					

(Continued)

Design team number (Continued)

Design process activities/strategies	Strongly agree	Agree	Do not know	Disagree	Strongly disagree
62. We did not adopt quick testing and instead we tested aggregately the software subsystem at the end of the project. What is the approximate percentage you assigned?					
63. We adopted effective programming techniques, such as selective switch case rather than enumerated switch case in the software programming code implementation, to minimise waste in the design development time.					
64. Our software designers preferred object oriented programming code implementation to structured programming code implementation because of the effectiveness of the former to achieve a better design because functions and subroutines are less effective as to reusability, scalability and manageability.					
65. Our software designers aim at implementing explanatory annotation of the software code for purposes of software code debugging and maintenance.					

(Continued)

Design team number (Continued)

Design process activities/strategies	Strongly agree	Agree	Do not know	Disagree	Strongly disagree
66. We adopted a parameterisation design strategy, that is, changing the value of one parameter changes the values of many related parameters accordingly, to facilitate software code debugging.[e]					

[a] We set an error-proof operational design management strategy for the design project. This point is intended to explore whether your design team considered early in the design process a strategy of eliminating the sources of mistakes in the design process as much as possible, for example, by constructing a three-dimensional computer-aided design (CAD) model (using Pro-Engineer) of the system assembly drawing, to minimise waste of time and resources due to avoidable mistakes.

[b] We adopted both sequential and concurrent design activities according to the nature of each task and the preceding relationships between them, to maximise utilisation of resources and shorten the design development time. This point is intended to address whether you took the advantage of planning the design activities such that some design activities are planned to be conducted at the same time (concurrently) as long as there is no constraint that would hinder this.

[c] Our design team estimated the duration of each project activity based on work breakdown structure, experience and comparison with comparable projects. This point is intended to explore how you estimated, in the planning phase of your design project, the duration of each design activity.

[d] Our mechanical subteam determined the optimal number of degrees of freedom of the system, for example, of picking-up and unloading mechanisms, that fulfil the functional requirements and minimise the number of needed controllable mechanisms. This point is intended to explore whether your design team thought early in the design process about and considered the degrees of freedom of movement of the mechanical design concept, for example, how many constraints are imposed on your mechanical design concept. Also, it explores whether your mechanical design subteam sought the minimum number of mechanical components that can achieve the design problem requirements.

[e] We adopted a parameterisation design strategy, that is, changing the value of one parameter changes the values of many related parameters accordingly, to facilitate software code debugging. This point is intended to explore whether your design team in the software subsystem considered the software code parameterisation strategy in the code architecture. By adopting this strategy, changing the values of specific parameters in the software code changes intentionally values of many related parameters accordingly. This strategy facilitates code debugging, code maintenance and code scalability.

Appendix B: Design management research questionnaire on the structural design project

The aim of this questionnaire is to explore, for research purposes, the hypothesised correlation and causality between design process activities and strategies on the one hand and the quality of the final product performance on the other. To explore this hypothesised correlation, we would like to collect some data from designers about their real design process activities and strategies. Consequently, we would like you to kindly answer this questionnaire. Thank you in advance for your cooperation.

Please, tick (✓) the **most appropriate evaluation category** to what your design team actually adopted and conducted in the design process.

Design team number

Design process activities/strategies	Strongly agree	Agree	Do not know	Disagree	Strongly disagree
2. We aimed at the simplest design rather than the cheapest and the lightest design.					
21. We spent more time on the conceptual and preliminary design phases than on the detailed design phase, to improve our opportunity for achieving a better design.					
22. We visited similar designs that solved similar design problems, to improve our opportunity to achieve a better design.					
23. We adopted cross fertilisation of design concepts among the design team, such as sharing ideas and/or modifying each other's ideas.					
28. We made more iterations in the conceptual and preliminary design phases than in the detailed design phase, to improve our opportunity of ending up with a better design.					
30. We adopted live meetings of the design team in the conceptual and preliminary design phases, and e-mails in the detailed design phase, as the principal communication method, to strike a balance between the synergy of face-to-face meetings on one hand, and the time effectiveness and message delivery effectiveness of the electronic means of communication, such as e-mails, on the other.					
47. We constructed a to-scale cardboard prototype, or mock-up, at the preliminary design phase, to improve our opportunity for achieving a better design.					

(Continued)

Design team number (Continued)

Design process activities/strategies	Strongly agree	Agree	Do not know	Disagree	Strongly disagree
49. We made a prediction of the failure modes and progressive failure of the design, for example, through drawing sketches to expect 2D in-plane failure and/or 3D out-of-plane failure, to improve our opportunity to achieve a better design.					
50. We made sketches and/or cardboard mock-ups to generate concepts in the conceptual design phase, to improve our opportunity of achieving a better design.					
53. We considered manufacturability in the conceptual design phase, to improve our opportunity for achieving a better design.					
54. We checked the accuracy of manufacturing and assembly of the final prototype, to avoid unexpected failure due to manufacturing defects and/or assembly mistakes.					

Appendix C: Design activities and strategies

1-STRA	Shifting complexity
2-STRA	Simplest design
3-ACT	Choosing the mechanical design concept first
4-STRA	Electronic distinguishing
5-STRA	Only standard components
6-STRA	Modular design
7-ACT	Systematic design approach
8-ACT	Quality of performing systematic design approach
9-STRA	Error proof
10-STRA	Tolerant design
11-STRA	Maximising functionality and minimising iterations
12-ACT	Top-down system structure decomposition
13-STRA	Fast response versus stability, accuracy and payload
14-STRA	Considering reliability
15-STRA	Considering maintainability
16-STRA	Considering safety
17-STRA	Error recovery
18-ACT	Largest number of iterations in software
19-ACT	Early verification
20-ACT	Software team leader
21-ACT	More time on conceptual and preliminary phases
22-ACT	Reviewing similar designs
23-ACT	Cross fertilisation of concepts among the team
24-ACT	Time to aggregate plan
25-ACT	Multidisciplinary team
26-STRA	Resource provisions
27-STRA	Modular testing
28-ACT	More iterations in conceptual and preliminary phases

29-ACT	Modular testable interdeliverables
30-ACT	Live meetings of the team in the conceptual and preliminary phases
31-STRA	Decentralised decision making
32-STRA	Sequential and concurrent design tasks
33-STRA	More interested in meeting delivery deadlines than cost
34-ACT	Estimating design activity duration
35-STRA	Redeployment of human resources
36-ACT	Modular deliverables and testing
37-ACT	Coordinating concerning the interrelated elements
38-ACT	Conducting design research
39-ACT	Starting the design process as early as possible and assigning time for presenting outcome
40-STRA	Considering critical path rather than critical chain
41-ACT	Documenting the outcome of design discussions
42-STRA	Having experienced designers to conceive assumptions of similar designs
43-STRA	Having experienced designers to prevent avoidable mistakes
44-ACT	Meeting collectively in conceptual rather than in detailed design
45-ACT	Meeting collectively in conceptual to choose the mechanical and electronics concepts
46-ACT	Conducting initial test experiments in the preliminary design phase
47-ACT	Constructing to-scale cardboard prototype
48-ACT	Strength-adaptable mechanical chassis
49-ACT	Predicting design progressive failure
50-ACT	Making sketches to generate concepts in conceptual design
51-ACT	Making CAD models in preliminary design
52-ACT	Building to-scale prototype to be tested in real operational environment
53-STRA	Considering manufacturability starting from conceptual design
54-ACT	Checking accuracy of manufacturing
55-ACT	Determining optimal number of degrees of freedom of the system
56-STRA	Error tracing
57-ACT	Extending sensor's robustness through software manipulation
58-ACT	Minimising number of reads and writes and maximising utilisation of channeling and port addressing
59-ACT	Functional analysis and software architecture
60-ACT	Quick testing of interdeliverables between modules
61-ACT	Having time for overall testing

62-ACT	Testing aggregately the software subsystem at the end of the project rather than adopting quick testing
63-ACT	Implementing effective programming techniques
64-STRA	Implementing object-oriented programming
65-ACT	Implementing software code explanatory annotation
66-STRA	Implementing software code parameterisation

References

ABB Robotics Product Guide. http://www.abb.com/product/ap/seitp327/cc4949febe7dcfe9c12573fa0057007a.aspx (Accessed 20 September 2013).

Abbas, E., Czwakiel, A., Valle, R., Ludlow, G. and Shah, S. (2011). "The practice of sustainable facilities management: Design sentiments and the knowledge chasm." *Architectural Engineering and Design Management*, June: 91–102.

Alder, H. L. and Roessler, E. B. (1962). *Introduction to Probability and Statistics*. New York: W. H. Freeman and Company.

Amano, T. (2014). "Prototyping in business model innovation: Exploring the role of design thinking in business model development." *Proceedings from the 19th DMI: Academic Design Management Conference*, London, UK, September 2–4, 2014.

Amon, C. H., Finger, S., Siewiorek, D. P. and Smailagic, A. (1995). "Integration of design education, research and practice at Carnegie Mellon University: A multi-disciplinary course in wearable computer design." *Frontiers in Education Conference Proceedings*, IEEE, 2: 14–22.

Andrew, J. P. and Sirkin, H. L. (2003). "Innovation for cash." *Harvard Business Review*, 81(9): 76–83.

Ascher, H. (2007). "Different insights for improving part and system reliability obtained from exactly same DFOM failure numbers." *Reliability Engineering and System Safety*, 92: 552–559.

Ashwini Nand, A., Singh, P. J. and Bhattacharya, A. (2014). "Do innovative organisations compete on single or multiple operational capabilities?" *International Journal of Innovation Management*, 18(3): 1–17.

Babbar, S., Behara, R. and White, E. (2002). "Mapping product usability." *International Journal of Operations & Production Management*, 22(10): 1071–1089.

Badding, S., Leigh, K. and Williams, A. (2014). "Models of thinking: Assessing the components of the design thinking process." *Proceedings from the 19th DMI: Academic Design Management Conference*, London, UK, September 2–4, 2014.

Bakerjian, R., Wick, C. and Benedict, J. T. (1993). "Tool and Manufacturing Engineers Handbook," Vol. VII: *Continuous Improvement*. Dearborn, MI: Society of Manufacturing Engineers.

Baxter, M. R. (1995) *Product Design: Practical Methods for the Systematic Development of New Products*. New York: Chapman & Hall.

Bayer, J. (2004). "Design for quality." In *5th International Workshop on Software Product-Family Engineering*, pp. 370–380. Berlin: Springer-Verlag, Siena, Italy, November 4–6, 2003.

Bayus, B. L. (1997). "Speed-to-market and new product performance trade-offs." *Journal of Product Innovation Management*, 14: 485–497.

Beesley, A. (1996). "Time compression in the supply chain." *Industrial Management & Data Systems*, 96(2): 12–16.

Beiter, K., Yang, T. and Ishii, K. (2006). "Preliminary design of amorphous products." In *Proceedings of the ASME Design Engineering Technical Conference*, Philadelphia, September 2006.

Belassi, W. and Tukel, O. I. (1996). "A new framework for determining critical success/failure factors in projects." *International Journal of Project Management*, 14(3): 141–151.

Benker, A. and Raduma, W. (2014). "Collaborative evaluation of design concepts." In *Proceedings From the 19th DMI: Academic Design Management Conference*, London, UK, September 2–4, 2014.

Bernard, R. (1999). *Early Evaluation of Product Properties Within the Integrated Product Development*. Aachen, Germany: Shaker Verlag.

Bigliardi, B. and Galati, F. (2014). "The implementation of TQM in R&D environments." *Journal of Technology Management & Innovation*, 9(2): 157–171.

Birmingham, R., Cleland, G., Driver, R. and Maffin, D. (1997). *Understanding Engineering Design: Context, Theory and Practice*. Upper Saddle River, NJ: Prentice Hall.

Blessing, L. T. M. (1994). *A Process-Based Approach to Computer-Supported Engineering Design*. PhD thesis, University of Twente, The Netherlands.

Blessing, L. T. M., Charkrabarti, A. and Wallace, K. M. (1995). "A design research methodology." In *Proceedings of the 10th International Conference on Engineering Design (ICED'95)*, Prague, Czech Republic, Vol. 1, pp. 50–55.

Boothroyd, G., Dewhurst, P. and Knight, W. A. (1991). "Selection of materials and processes for component parts." In *Proceedings of the 1992 NSF Design and Manufacturing Systems Conference*, Atlanta, GA, January 1992, pp. 255–263.

Browning, T. R. (1998). "Sources of performance risk in complex system development." MIT Lean Aerospace Initiative, pp. 1–8.

Browning, T. R. (2001). "Applying the design structure matrix to system decomposition and integration problems: A review and new directions." *IEEE Transactions on Engineering Management*, 48(3): 292–306.

Browning, T. R. and Eppinger, S. D. (2002). "Modeling the impact of process architecture on cost and schedule risk in product development." *IEEE Transactions on Engineering Management*, 49(4): 428–442.

Browning, T. R. and Ramasesh, R. V. (2007). "A survey of activity network-based process models for managing product development projects." *Production and Operations Management*, 16(2): 217–240.

Browning, T. R., Deyst, J. J. and Eppinger, S. D. (2002). "Adding value in product development by creating information and reducing risk." *IEEE Transactions on Engineering Management*, 49(4): 443–458.

Browning, T. R., Fricke, E. and Negele, H. (2005). *Key Concepts in Modelling Product Development Processes*. Hoboken, NJ: Wiley Interscience, pp. 104–128.

Calantone, R. J. and Di Benedetto, C. A. (2000). "Performance and time to market: Accelerating cycle time with overlapping stages." *IEEE Transactions on Engineering Management*, 47(2): 232–244.

Cambridge University, Engineering Department (CUED). (2007a). *First Year Undergraduate Structural Design Project*. Cambridge, UK: Cambridge University Press.

Cambridge University, Engineering Department (CUED). (2007b). *Second Year Undergraduate Integrated Design Project.* Cambridge, UK: Cambridge University Press.

Castillo, O., Trujillo, L. and Melin, P. (2007). "Multiple objective genetic algorithms for path-planning optimization in autonomous mobile robots." *Soft Computing*, 11(3): 269–279.

Chalupnik, M. J., Eckert, C. M. and Clarkson, P. J. (2006). "Modelling design processes to improve robustness." In *6th Integrated Product Development Workshop, IPD 2006*, Schonebeck/Bad Salzelmen b. Magdeburg, Germany.

Chalupnik, M. J., Wynn, D. C., Eckert, C. M. and Clarkson, P. J. (2007). "Understanding design process robustness: A modelling approach." In *16th International Conference on Engineering Design (ICED'07)*, Paris, pp. 455–456.

Chalupnik, M. J., Wynn, D. C., Eckert, C. M. and Clarkson, P. J. (2008). "Analysing the relationship between design process composition and robustness to task delays." In *International Design Conference (Design 2008)*, Dubrovnik, Croatia.

Chan, A. P. C., Scott, D. and Chan, A. P. L. (2004). "Factors affecting the success of a construction project." *Journal of Construction Engineering and Management*, January/February: 153–155.

Chan, A. P. C., Scott, D. and Lam, E. W. M. (2002). "Framework of success criteria for design/build projects," *Journal of Management in Engineering*, 18(3): 120–128.

Chao, L. P., Tumer, I. and Ishii, K. (2005). "Design process error-proofing: Benchmarking the NASA development life-cycle." In *IEEE Aerospace Conference*, Big Sky, MT, March 5–12, 2005, pp. 4327–4338.

Chen, D. and Cheng, F. (2000). "Integration of product and process development using rapid prototyping and work-cell simulation technologies." *Journal of Industrial Technology*, 16(1): 2–5.

Chen, W. and Allen, J. K. (1996). "A procedure for robust design: Minimizing variations caused by noise factors and control factors." *Transaction of the ASME Journal, Journal of Mechanical Design*, 118(4): 478–485.

Chen, J., Reilly, R. R., Lynn, G. S. (2005). "The Impacts of Speed-to-Market on New Product Success: The Moderating Effects of Uncertainty," *IEEE Transactions on Engineering Management*, 52(2).

Christensen, H. I., Dillmann, R., Hägele, M., Kazi, A. and Norefors, U. (2008). "European robotics," *European Robotics Forum*, July 2008.

Clark, K. B. and Baldwin, C. Y. (2000). *Design Rules, Vol. 1: The Power of Modularity.* Cambridge, MA: MIT Press.

Clark, K. B. and Fujimoto, T. (1991). *Product Development Performance: Strategy, Organization, and Management in the World Auto Industry.* Boston: Harvard Business School Press.

Clark, K. B. and Fujimoto, T. (1992). "Product development and competitiveness." *Journal of the Japanese and International Economies*, 6: 101–143.

Clarkson, J. P. and Eckert, C. (2005). *Design Process Improvement: A Review of Current Practice.* New York: Springer Science+Business Media.

Clarkson, P. J., Melo, A. and Connor, A. (2000). "Signposting for design process improvement." In *Proceedings of Artificial Intelligence in Design*, Worcester, USA, June 26–29, 2000, pp. 333–353.

Clausing, D. P. (1994). *Total Quality Development: A Step-by-Step Guide to World Class Concurrent Engineering.* New York: American Society of Mechanical Engineers.

Cohen, J. (1988). *Statistical Power Analysis for the Behavioural Sciences.* New York: Academic Press.

Cohen, M. A., Eliashberg, J. and Ho, T. (1996). "New product development: The performance and time-to-market trade-off." *Journal of Strategic Research (JSTOR)*, 42(2): 173–186.

Cook, T. D. and Campbell, D. T. (1979a). "*Quasi-experimentation: Design and Analysis Issues for Field Settings*," Boston: Houghton-Mifflin.

Cook, T. D. and Campbell, D. T. (1979b). "Four kinds of validity." In R. T. Mowday and R. M. Steers (Eds.), *Research in Organizations: Issues and Controversies*. Santa Monica, CA: Goodyear Publishing.

Cooper, R. and Slagmulder, R. (1997). *Target Costing and Value Engineering*. New York: Productivity Press.

Cooper, R. G. (1994). "Third generation new product processes." *Journal of Product Innovation Management*, 11: 3–14.

Coulibaly, A., Houssin, R. and Mutel, B. (2008). "Maintainability and safety indicators at design stage for mechanical products." *Computers in Industry*, 59(5): 438–449.

Court, A. W. (1998). "Issues for integrating knowledge in new product development: Reflections from an empirical study." *Knowledge-Based Systems*, 11: 391–398.

Cristiano, J. J., Liker, K. J. and White, C. C. (2000). "Customer-driven product development through quality function deployment in the U.S. and Japan." *Journal of Production Innovation Management*, 17: 286–308.

Cross, N. (1984). *Developments in Design Methodology*. Chichester: John Wiley & Sons.

Cross, N. (1994). *Engineering Design Methods: Strategies for Product Design*. Chichester: John Wiley & Sons.

Cross, N. (2008). *Engineering design methods: Strategies for product Design*. Wiley, ISBN-10:0470519266.

Dahl, D. W., Chattopadhyay, A. and Gorn, G. J. (2001). "The importance of visualisation in concept design." *Design Studies*, 22(1): 5–26.

Debruyne, M., Rudy, M., Griffin, A., Hart, S., Hultink, E. J. and Robben, H. (2002). "The impact of new product launch strategies on competitive reaction in industrial markets." *Journal of Product Innovation Management*, 19(2): 159–170.

De Zubielqui, G. C., Lindsay, N. J. and Connor, A. O. (2014). "How product, operations, and marketing sources of ideas influence innovation and entrepreneurial performance in Australian SMEs." *International Journal of Innovation Management*, 18(2): 1450017-1–1450017-25.

Doherty, R., Wrigley, C., Matthews, J. and Bucolo, S. (2014). "Climbing the design ladder: Step by step." In *Proceedings from the 19th DMI: Academic Design Management Conference*, London, UK, 2014.

Driva, H. and Pawar, K. S. (1999). "Performance measurement for product design and development in a manufacturing environment." *International Journal of Production Economics*, 60: 61–68.

Drucker, P. F. (2002). "The discipline of innovation." *Harvard Business Review*, 80(8): 95–100.

Eckert, C., Clarkson, J. P. and Stacey, M. K. (2003). "The spiral of applied research: A methodological view on integrated design research." In *Proceedings of 14th International Conference on Engineering Design (ICED'03)*, Stockholm, Sweden, pp. 245–246.

Eckert, C. M., Stacey, M. K. and Clarkson, P. J. (2004). "The lure of the measurable in design research." In *Design 2004*, Dubrovnik, Croatia, pp. 21–26.

Eisenhardt, K. M. and Tabrizi, B. N. (1995). "Accelerating adaptive processes: Product innovation in the global computer industry." *Administrative Science Quarterly*, 40: 84–110.

Elmoselhy, S. A. M. (2014). "Mobile robots dsign guideline based on an empirical study of the mobile robots design process." *International Review of Mechanical Engineering*, 8(3): 489–494.

Elmoselhy, S. A. M. (2015). "Empirical investigation of a hybrid lean-agile design paradigm for mobile robots." *Journal of Intelligent Systems*, 24(1).

Elmualim, A. and Gilder, J. (2013). "BIM: Innovation in design management, influence and challenges of implementation." *Architectural Engineering and Design Management*, August: 183–199.

Emmitt, S. (2011). "Lean design management." *Architectural Engineering and Design Management*, June: 67–69.

Eppinger, S. D. (2001). "Innovation at the speed of information." *Harvard Business Review*, 79(1): 149–158.

Erhun, F. (2007). "The art of managing new product transitions." *MIT Sloan Management Review*, 48(3): 73–80.

Eris, O. (2004). *Effective Inquiry for Innovative Engineering Design: From Basic Principles to Applications*. New York: Springer Science+Business Media.

Esterman, M., Gerst, P., Stiebitz, P. H. and Sihii, K. (2005). "A framework for warranty prediction during product development." In *Proceedings of IMECE 2005*, Orlando, FL, November 5–11, 2005.

Evans, J. H. (1959). "Basic design concepts." *Journal of the American Society of Naval Engineers*, November: 671–678.

Fanuc Robotics, M410 ib series, Product Guide. http://www.fanuc.co.jp/en/product/catalog/pdf/M-410iB(E)_v07_s.pdf (Accessed 20 September 2013).

Festinger, L. and Katz, D. (1996). *Research Methods in the Behavioural Sciences*. New York: Holt, Rinehart and Winston.

Finger, S. and Dixon, J. R. (1989). "A review of research in mechanical engineering design, Part I: Descriptive, prescriptive and computer-based models of design processes." *Research in Engineering Design*, 1(1): 51–68.

Ford, D. N. and Sterman, J. D. (1998). "Dynamic modeling of product development processes." *System Dynamics Review*, 14(1): 31–68.

French, M. J. (1999). *Conceptual Design for Engineers*. New York: Springer Science+Business Media.

Gardien, P., Deckers, E. and Christiaansen, G. (2014). "Innovating innovation – Deliver meaningful experiences in ecosystems." In *Proceedings from the 19th DMI: Academic Design Management Conference*, London, UK, September 2–4, 2014.

Gemser, G., Jacobs, D. and Ten Cate, R. (2006). "Design and competitive advantage in technology-driven sectors: The role of usability and aesthetics in Dutch IT companies." *Technology Analysis and Strategic Management*, 18(5): 561–580.

Gershenson, J. K. and Prasad, G. J. (1997). "Modularity in product design for manufacturability." *International Journal of Agile Manufacturing*, 1(1): 99–110.

Goh, T. N. (1993). "Taguchi methods: Some technical, cultural and pedagogical Perspectives," *Quality and Reliability Engineering International*, 9(3): 185–202.

Gonzalez De Santos, P., Garcia, E. and Estremera, J. (2007). "Improving walking-robot performances by optimizing leg distribution." *Autonomous Robots*, 23(4): 247–258.

Govindarajan, S. (1993). *Strategic Cost Management: The New Tool for Competitive Advantage*. New York: Free Press.

Graner, M. and Mißler-behr, M. (2014). "Method application in new product development and the impact on cross-functional collaboration and new product success." *International Journal of Innovation Management*, 18(1). doi: 10.1142/S1363919614500029.

Green, K. and Raman, R. (2014). "Innovation hit rate, product advantage, innovativeness, and firm performance." *International Journal of Innovation Management*, 18(5). doi: 10.1142/S1363919614500388.

Guenov, M. D. and Barker, S. G. (2005). "Application of axiomatic design and design structure matrix to the decomposition of engineering systems." *Systems Engineering*, 8(1): 29–40.

Hales, C. (2004). *Managing Engineering Design*. New York: Springer Science+Business Media.

Hall, A. D. (1962). *A Methodology for Systems Engineering*. New York: Van Nostrand Reinhold.

Hamada, T. (2014). "Adaptation to technological innovation and corporate core-rigidities." *International Journal of Innovation Management*, 18(2): 1450019-1–1450019-21.

Hansen, G. K. and Olsson, N. O. E. (2011). "Layered project–layered process: Lean thinking and flexible solutions." *Architectural Engineering and Design Management*, June: 70–84.

Hardt, M., Stryk, O., Wollherr, D. and Buss, M. (2000). "Design of an autonomous fast-walking humanoid robot." In *International Conference on Climbing and Walking Robots*, Paris, pp. 391–398, September 2000.

Harkins, R., Ward, J., Vaidyanathan, R., Boxerbaum, A. S. and Quinn, R. D. (2005). "Design of an autonomous amphibious robot for surf zone operations: Part II. Hardware, control implementation and simulation." In *Proceedings, 2005 IEEE/ASME International Conference on Advanced Intelligent Mechatronics*, Monterey, CA, pp. 1465–1470, July 2005.

Hastings, D. and McManus, H. (2004). "A framework for understanding uncertainty and its mitigation and exploitation in complex systems." In *Proceedings of 2004 Engineering Systems Symposium*, MIT, MA, March 29–31, 2004, pp. 1–19.

Huang, C. (2000). "Overview of modular product development." *Proceedings of National Science Council, ROC(A)*, 24(3): 149–165.

Huang, G. O. (1996). *Design for X: Concurrent Engineering Imperatives*. New York: Springer Science+Business Media.

Huang, P., Pan, S. L. and Ouyang, T. H. (2014). "Developing information processing capability for operational agility: Implications from a Chinese manufacturer." *European Journal of Information Systems*, 23(4): 462–480.

Hundal, M. (1997). *Systematic Mechanical Designing: A Cost and Management Perspective: Cost-Based Mechanical Design & Product Development*. New York: American Society of Mechanical Engineers.

Inauen, M. and Schenker-Wicki, A. (2012). "Fostering radical innovations with open innovation." *European Journal of Innovation Management*, 15(2): 212–231.

International Federation of Robotics. (2006). "Executive summary of 2005 world robot market."

Jenkins, J. and Fife, T. (2014). "Designing for disruption: Strategic business model innovation." *Proceedings from the 19th DMI: Academic Design Management Conference*, London, UK, September 2–4, 2014.

Jenkins, J. and Golsby-Smith, T. (2013). "(Im)proving it: Designing a measurement system that nourishes innovation." *Design Management Review*, 24(4): 40–46.

Jerrard, R. N., Barnes, N. and Reid, A. (2008). "Design, risk and new product development in five small creative companies." *International Journal of Design*, 2(1): 21–30.

Jikar, V. K. and Ragsdell, K. M. (2005). "Rapid product development." In *26th ASEM National Conference Proceedings, Organizational Transformation: Opportunities and Challenges*, Virginia Beach, VA, pp. 544–550, October 2005.

Joglekar, N. and Ford, D. N. (2005). "Product development resource allocation with foresight." *European Journal of Operational Research*, 160(1): 72–87.

Johnson, R. A. (1976). *Management, Systems, and Society: An Introduction*. Santa Monica, CA: Goodyear Publishing.

Keates, S. and Clarkson, P. J. (2003). *Countering Design Exclusion: An Introduction to Inclusive Design*. New York: Springer Science+Business Media.

Khalaf, F. and Yang, K. (2006). "Product development processes – from deterministic to probabilistic: A design for 6-sigma approach to lean product validation, Part II." *International Journal of Product Development*, 3(1): 18–36.

Klaus, E., Kiewert, A. and Lindemann, U. (2007). *Cost-Effective Design*. Berlin: Springer-Verlag.

Krejcie, R. and Morgan, D. (1970). "Determining sample size for research activities." *Educational and Psychological Measurement*, 30: 607–610.

Krishnan, V. (1998). "Modeling ordered decision making in product development." *European Journal of Operational Research*, 111(2): 351–368.

Krishnan, V. and Ulrich, K. T. (2001). "Product development decisions: A review of the literature." *Management Science*, 47(1): 1–21.

Leach, L. P. (2000). *Critical Chain Project Management Improves Project Performance*. Advanced Projects Institute, Idaho Falls, ID.

Lévárdy, V. and Browning, T. R. (2005). *Adaptive Test Process – Designing a Project Plan that Adapts to the State of a Project*. San Diego: INCOSE Publications.

Lindemann, U. (2003). *Human Behaviour in Design: Individuals, Teams, Tools*. Berlin: Springer-Verlag.

Liou, F., Slattery, K., Kinsella, M., Newkirk, J., Chou, H. and Landers, R. (2007). "Applications of a hybrid manufacturing process for fabrication of metallic structures." *Rapid Prototyping Journal*, 13(4): 236–244.

Liu, J. and Su, J. (2014). "Market orientation, technology orientation and product innovation success: Insights from CoPS." *International Journal of Innovation Management*, 18(4). doi: 10.1142/S1363919614500200.

Loughran, N. and Rashid, A. (2004). "Managing variability throughout the software development lifecycle." Computing Department, Lancaster University, pp. 1–3.

Mahmoud-Jouini, S. B., Midler, C. and Garel, G. (2004). "Time-to-market vs. time-to-delivery: Managing speed in engineering, procurement, and construction projects." *International Journal of Project Management*, 22: 359–367.

Maier, A. M., Kreimeyer, M., Hepperle, C., Eckert, C. M., Lindemann, U. and Clarkson, P. J. (2008). "Exploration of correlations between factors influencing communication in complex product development." *Concurrent Engineering*, 16(1): 37–59.

Maier, J. F., Wynn, D. C., Biedermann, W., Lindemann, U. and Clarkson, P. J. (2014). "Simulating progressive iteration, rework and change propagation to prioritise design tasks." *Research in Engineering Design*, 25: 283–307.

Maier, J. R. A., Ezhilan, T. and Fadel, G. M. (2007). "The affordance structure matrix – A concept exploration and attention directing tool for affordance based design." *Proceedings of ASME Design Theory and Methodology Conference,* ASME IDETC/DTM, DETC2007-34526, Las Vegas, NV, 2007.

Manktelow, K. I. (1999). "*Reasoning and Thinking.* New York: Psychology Press.

McMath, M. R. and Forbes, T. (1998). *What Were They Thinking?* New York: Times Business Random House.

Nishiguchi, T. (1996). *Managing Product Development.* Oxford: Oxford University Press.

O'Dell, T. H. (1988). *Electronic Circuit Design.* Cambridge, UK: Cambridge University Press.

O'Donovan, B. D., Clarkson, P. J. and Melo, A. F. (2002). "Development process simulation using 'Signposting'." In *Concurrent Engineering Conference 2002 (CECONF 2002),* Cranfield University, UK.

Otto, K. and Antonsson, E. (1991). "Trade-off strategies in engineering design." *Research in Engineering Design,* 3(2): 87–104.

Owen, R., Amor, R., Palmer, M., Dickinson, J., Tatum, C. B., Kazi, A., Prins, M., Kiviniemi, A. and East, B. (2011). "Challenges for integrated design and delivery solutions." *Architectural Engineering and Design Management,* June: 232–240.

Pahl, G. and Beitz, W. (1998). *Engineering Design: A Systematic Approach.* Berlin: Springer-Verlag.

Paladino, A. (2007). "Investigating the drivers of innovation and new product success: A comparison of strategic orientations." *Journal of Product Innovation Management,* 24(6): 534–553.

Pearce, R. D. (1999). "Decentralised R&D and strategic competitiveness: Globalised approaches to generation and use of technology in multinational enterprises MNEs." *Research Policy,* 28: 157–178.

Prabhakar Murthy, D. N., Rausand, M. and Østerås, T. (2008). *Product Reliability: Specification and Performance.* Berlin: Springer-Verlag.

Pugh, S. (1991a). *Total Design: Integrated Methods for Successful Product Engineering.* Reading, MA: Addison-Wesley.

Pugh, S. (1991b). "Concept selection and design vulnerability." In *Proceedings from Design Productivity Institute Conference,* Honolulu, February 1991.

Pyzdek, T. and Keller, P. A. (2003). *Quality Engineering Handbook,* CRC Press.

Repenning, N. and Sterman, J. (2001). "Nobody ever gets credit for fixing problems that never happened: Creating and sustaining process improvement." *California Management Review,* 43(4): 64–88.

Rob, M. A. (2004). "Issues of structured vs. object-oriented methodology of systems analysis and design." *Issues in Information Systems,* 5(1): 275–280.

Robson, C. (2002). *Real World Research.* Boston: Blackwell Publishing.

Rosenhead, J. (1980). "Planning under uncertainty: 2. A methodology for robustness analysis." *Journal of the Operational Research Society,* 31(4): 331–341.

Rostami, V., Sojodishijani, O., Ebrahimijam, S. and Mohsenizanjani Nejad, A. (2005). "Fuzzy error recovery in feedback control for three wheel omnidirectional soccer robot." In *Proceedings of World Academy of Science, Engineering and Technology,* Vol. 9, pp. 91–94, November 2005.

Rust, R. T., Zahorik, A. J. and Keiningham, T. L. (1995). "Return on quality (ROQ): Making service quality financially accountable." *Journal of Marketing,* 59: 58–70.

Sandstr, J. and Toivanen, J. (2002). "The problem of managing product development engineers: Can the balanced scorecard be an answer?" *International Journal Production Economics*, 78: 79–90.

Santos, F. P. and Morillo, M. (2014). "Materiality, design and brand management." *Proceedings from the the 19th DMI: Academic Design Management Conference*, London, UK, September 2–4, 2014.

Schneider, S. and Spieth, P. (2014). "Business model innovation and strategic flexibility: Insights from an experimental research design." *International Journal of Innovation Management*, 18(6): 1–21.

Seidel, V. P. and Fixson, S. K. (2013). "Adopting design thinking in novice multidisciplinary teams: The application and limits of design methods and reflexive practices." *Journal of Product Innovation Management*, 30(Suppl. S1): 19–33.

Sekaran, U. (2003). *Research Methods for Business*. Hoboken, NJ: John Wiley & Sons.

Shenhar, A. J. and Wideman, R. M. (1996). "Project Management: From Genesis to Content to Classification," paper presented at Operations Research and Management Science (INFORMS), Washington, DC.

Simeone, L. (2014). "Interplay between UCD and design management in creating an interactive platform to support low carbon economy." In *Proceedings from the 19th DMI: Academic Design Management Conference*, London, UK, September 2–4, 2014.

Sivaloganathan, S., Shahin, T. M. M., Cross, M. and Lawrence, M. (2000). "A hybrid systematic and conventional approach for the design and development of a product: A case study." *Design Studies*, 21(1): 59–74.

Slater, S. F. and Mohr, J. J. (2006). "Successful development and commercialization of technological innovation: Insights based on strategy type." *Journal of Product Innovation Management*, 23(1): 26–33.

Smith, P. G. and Reinertsen, D. G. (1997). *Developing Products in Half the Time: New Rules, New Tools*. New York: John Wiley & Sons.

Smith, R. P. and Eppinger, S. D. (1997a). "A predictive model of sequential iteration in engineering design." *Management Science*, 43(8): 1104–1120.

Smith, R. P. and Eppinger, S. D. (1997b). "Identifying controlling features of engineering design iterations." *Management Science*, 43(3): 276–293.

Sola, D., Scalabrini, G. and Scarso Borioli, G. (2014). "Reducing uncertainty through disciplined experimentation." In *Proceedings from the 19th DMI: Academic Design Management Conference*, London, UK, September 2–4, 2014.

Sommerville, I. (2007). *Software Engineering*. Upper Saddle River, NJ: Pearson Education.

Sosa, M. E., Eppinger, S. D. and Rowles, C. R. (2004). "The misalignment of product architecture and organizational structure in complex product development." *Management Science*, 50(12): 1674–1689.

Soy, S. (1997). "The case study as a research method." *Uses and Users of Information*, Vol. LIS391D.1, Spring, pp. 1–7. Retrieved January 16, 2003, http://www.gslis.utexas.edu/ssoy/usesusers/L391d1b.htm.

Stevenson, W. J. (2002). *Operations Management*. New York: McGraw-Hill.

Straker, K. and Wrigley, C. (2014). "Design innovation catalyst tools to facilitate organisational change." In *Proceedings from the 19th DMI: Academic Design Management Conference*, London, UK, September 2–4, 2014.

Susman, G. I. (1992). *Integrating Design and Manufacturing for Competitive Advantage*. Oxford: Oxford University Press.

Swan, K. S., Kotabe, M. and Allred, B. B. (2005). "Exploring robust design capabilities, their role in creating global products, and their relationship to firm performance." *The Journal of Product Innovation Management*, 22: 144–146.

Taguchi, G. and Clausing, D. (1990). "Robust quality." *Harvard Business Review*, January–February: 65–75.

Tatikonda, M. V. and Montoya-Weiss, M. M. (2001). "Integrating operations and marketing perspectives of product innovation: The influence of organizational process factors and capabilities on development performance." *Management Science*, 47(1): 151–172.

Tepic, M., Kemp, R., Omts, O. and Fortuin, F. (2013). "Complexities in innovation management in companies from the European industry: A path model of innovation project performance determinants." *European Journal of Innovation Management*, 16(4): 517–550.

Terninko, J., Zusman, A. and Zlotin, B. (1998). *Systematic Innovation: An Introduction to TRIZ (Theory of Inventive Problem Solving)*. Boca Raton, FL: CRC Press.

The British Standards Institution. (2008). "BS 7000–1:2008 Design management systems: Guide to managing innovation."

Theilmann, C. and Hukauf, M. (2014). "Customer integration in mass customisation: A key to corporate success." *International Journal of Innovation Management*, 18(3): 1440002-1–1440002-23.

Thurston, D. L. and Locascio, A. (1993). "Multiattribute design optimization and concurrent engineering." In H. Parsaei and W. Sullivan (Eds.), *Concurrent Engineering: Contemporary Issues and Modern Design Tools*. New York: Springer-Verlag.

Thurston, D. L. and Locascio, A. (1994). "Decision theory for design economics." *The Engineering Economist*, 40(1): 41–71.

Tomatis, N., Philippsen, R., Jensen, B., Arras, K. O., Terrien, G. Piguet, R. and Siegwart, R. (2002). *Building a Fully Autonomous Tour Guide Robot: Where Academic Research Meets Industry*. Autonomous Systems Lab, Swiss Federal Institute of Technology, Zurich.

Tomiyama, T. and Meijer, B. R. (2006). "Directions of next generation product development." In H. ElMaraghy and W. H. ElMaraghy (Eds.), *Advances in Design*. New York: Springer Science+Business Media, pp. 27–35.

Topping, P., Mclnroy, J., Livley, W. and Shepparc, S. (1987). "Express-rapid prototyping and product development via integrated knowledge-based executable specifications." Association for Computing Machinery *Annual Conference Proceedings of the 1987 Fall Joint Computer Conference on Exploring Technology: Today and Tomorrow*, IEEE Computer Society Press, Los Alamitos, CA, 1987.

Tribelsky, E. and Sacks, R. (2011). "An empirical study of information flows in multidisciplinary civil engineering design teams using lean measures." *Architectural Engineering and Design Management*, June: 85–101.

Ulrich, K. T. and Eppinger, S. D. (2003). *Product Design and Development*. New York: McGraw-Hill.

Valencia, A., Mugge, R., Schoormans, J. P. L. and Schifferstein, H. N. J. (2014). "Challenges in the design of smart product-service systems (PSSS): Experiences from practitioners." *Proceedings from the 19th DMI: Academic Design Management Conference*, London, UK, September 2–4, 2014.

Veryzer, R. W. (1998a). "Discontinuous innovation and the new product development process." *Journal of Product Innovation Management*, 15(4): 304–321.

Veryzer, R. W. (1998b). "Key factors affecting customer evaluation of discontinuous new products." *Journal of Product Innovation Management*, 15(2): 136–150,

Wackerly, D. D., Mendenhall, W. and Scheaffer, R. L. (1996). *Mathematical Statistics with Applications*. Pacific Grove, PA: Duxbury Press.

Wallace, K. and Clarkson, P .J. (1999). *Introduction to the Design Process*. Cambridge University Engineering Department (CUED). Cambridge, UK: Cambridge University Press.

Wasserman, A. I., Pircher, P. A. and Muller, R. J. (1990). "The object-oriented structured design notation for software design representation." *Computer, IEEE Computer Society*, 23(3): 50–63.

Wegner, P. (1984). "Capital-intensive software technology." *Software, IEEE*, 1(3): 7–10.

Wilde, D. J. (1992). "Product quality in optimization models." *Proceedings of the ASME 4th International Conference on Design Theory and Methodology*, Scottsdale, pp. 237–241, September 1992.

Wildman, G. (2014). "Live, actionable and tangible: Teaching design strategy." In *Proceedings from the 19th DMI: Academic Design Management Conference*, London, UK, September 2–4, 2014.

Wollherr, D. and Buss, M. (2001). "Cost oriented VR-simulation environment for computer aided control design." In *Proceedings of the 6th IFAC Symposium on Cost Oriented Automation*, Berlin, Germany, pp. 1–6.

Wynn, D. C. (2007). *Model-Based Approaches to Support Process Improvement in Complex Product Development*. PhD thesis, The University of Cambridge.

Wynn, D. C., Clarkson, P. J. and Eckert, C. M. (2005). "A model-based approach to improve planning practice in collaborative aerospace design." In *Proceedings of ASME 2005, International Design Engineering Technical Conferences and Computers and Information in Engineering Conference*, Paper # 200585297, pp. 1–12.

Wynn, D. C., Eckert, C. M. and Clarkson, P. J. (2006). "Applied Signposting: A modelling framework to support design process improvement." In *Proceedings of IDETC/CIE 2006*, Philadelphia, PA, September 10–13, 2006, Paper # 2006-99402, pp. 1–10.

Yassine, A., Joglekar, N., Braha, D., Eppinger, S. and Whitney, D. (2003). "Information hiding in product development: The design churn effect." *Research in Engineering Design*, 14: 145–161.

Yavuz, H. (2007). "An integrated approach to the conceptual design and development of an intelligent autonomous mobile robot." *Robotics and Autonomous Systems*, 55: 498–512.

Yin, R. K. (2002). *Case Study Research: Design and Methods*. Thousand Oaks, CA: SAGE Publications.

Index